Simplicius
On Aristotle's *Physics 7*

Simplicius
On Aristotle's *Physics 7*

Translated by
Charles Hagen

Cornell University Press
Ithaca, New York

For information address Cornell University Press,
Sage House, 512 East State Street, Ithaca, New York 14851-0250

First published 1994 by Cornell University Press.

Library of Congress Cataloging-in-Publication Data

Simplicius, of Cilicia.
On Aristotle's Physics 7 / Simplicius ; translated by Charles Hagen.
p. cm. – (Ancient commentators on Aristotle)
Includes bibliographical references and index.
ISBN 0-8014-2992-7
1. Aristotle. Physics. 2. Science, Ancient. 3. Physics–Early works to 1800. 4. Motion–Early works to 1800. 5. Knowledge, Theory of–Early works to 1800. I. Hagen, Charles. II. Title. III. Title: On Aristotle's Physics seven. IV. Series.
Q151.A8S563 1994
530–dc20 93-26377

Acknowledgments

The present translations have been made possible by generous and imaginative funding from the following sources: the National Endowment for the Humanities, Division of Research Programs, an independent federal agency of the USA; the Leverhulme Trust; the British Academy; the Jowett Copyright Trustees; the Royal Society (UK); Centro Internazionale A. Beltrame di Storia dello Spazio e del Tempo (Padua); Mario Mignucci; Liverpool University. The editor wishes to thank Robert Wardy, Catherine Atherton, A.R. Lacey and Peter Lautner for their comments on the translation and Ian Crystal, Paul Opperman and Dirk Baltzly for their help in preparing the volume for press.

Printed in Great Britain

Contents

Introduction

Richard Sorabji

Simplicius is important for the understanding of Aristotle's *Physics* Book 7.[1] He explains that the book seemed redundant to many because Book 8 covers the same subject, the case for a prime mover, with more exactitude. Accordingly, it had been left out of Eudemus' treatment of the *Physics*, and discussion of it had been skimped by Themistius. Simplicius did not necessarily have access to Philoponus' commentary on it, of which the Arabic summary is translated in this series.

Simplicius' response is that the book shows Aristotle's acumen and is connected with other books in the *Physics*, but that it was superseded by Book 8. It is, however, germane enough to have been included by some as a part of the *Physics*. Simplicius also reports the existence of the two versions of the Greek text, A and B. He himself makes more use of version A, but his references to B show that originally B extended at least into Chapter 4, whereas now it is extant only for the first three chapters. In Chapter 1, Aristotle argues that to avoid an infinity of movements, there must be a first moved mover (A 242b71-2; cf. B 242b34). Simplicius puts the point by saying there is a first mover not moved by anything else (1047,15-16; cf. 1042,13-14, 26-7; 1045,5; 1112,1-2). So far, both formulations leave open that it might be *self*-moved, as Plato believed the soul to be. But as Robert Wardy brings out in his excellent study of *Physics* 7,[2] self-motion is ruled out earlier in the chapter (A 241b39-44, 242a35-49; B 241b27-33, 242a1-13), particularly clearly in version B, which insists that in apparent self-movers what is moved is moved by something else (242a3,12-13). The unstated implication is that the first moved mover presupposes something else to move it, but that since it is itself the first thing to be moved, the presupposed mover must be unmoved.

Book 8 argues explicitly for an *unmoved* mover. But Book 7, Chapter 1, contributes an argument that is missing from Book 8. An

1 Simplicius *in Phys.* 1036,1-1037,10.

2 Robert Wardy, *The Chain of Change: A Study of Aristotle's Physics VII*, Cambridge 1990. This introduction draws heavily on Wardy and on the excellent notes supplied by Charles Hagen to the translation below.

infinite chain of moved movers is impossible, because their motions would sum into one infinite motion, and it has been proved earlier (*Physics* 6.7) that an infinite motion cannot be performed in a finite time (A 242a49-242b53; B 242a15-242b19). The motions would sum, because agent and patient have to be in contact with, or continuous with, each other (A 242b53-243a31; B 242b20-243a2).

Simplicius is very helpful in explaining how the other chapters of Book 7 fit into the first one. Chapter 2, he says (1048,1-6) defends the last assertion, that agent and patient must be in contact with each other, in all kinds of change, motion, change of size and change of quality, at least where that change is produced, by *sensible* qualities.

Chapter 3 argues, he points out (1061,25-9), that there is no other kind of qualitative change. Apparent counter-examples are not genuine cases of qualitative change, although (246a6-9; 246b14-17; 248a4) they may involve something *else* undergoing qualitative change. The point of this last concession is not explained by Aristotle. Simplicius takes it (p. 47 at note 229) as an explanation of why the apparent counter-examples seem to be cases of qualitative change. But Wardy is right that one could draw a conclusion more relevant to Aristotle's purpose: the counter-examples, at least indirectly, conform to the requirement that agent and patient be in contact.

Simplicius does not explain the relevance of Chapter 4 on the extent to which different types of change and motion can be compared in speed. Wardy suggests the purpose is to remove a threat to the procedure adopted in Chapter 1 of summing infinitely many motions into one infinite one. The chapter certainly shows Aristotle's acumen, when he discusses criteria for comparability of speed, including criteria for difference in meaning of terms such as 'fast'.

The last Chapter, 5, as Simplicius says (1102,29-1103,3), relates to the discussion in Chapter 4 of the comparability of motions. For it looks to see whether all motions are comparable in observing the same proportionalities. Do the force (*iskhus*), the weight moved, the distance covered and the time taken stand in fixed ratios, so that doubling one doubles or halves another? Aristotle finds a problem (250a12-19), for if you double the weight or halve the force, you may fall below a threshold and get no motion at all. If many haulers can move a ship, it does not follow that one hauler can move it a little. This also solves a paradox of Zeno's (250a19-25), who argued that if a falling bushel of millet grains makes a sound, so will any fraction of a grain. Aristotle's reply is more developed in *On Sense Perception* 6, 445b29-446a20. A part exists only potentially within a whole, and if a part is small enough it cannot exist separately from the whole either, since it would dissolve. Although such tiny parts are not detected within the whole (445b31; 446a2, 4, 15), they can still be called perceptible, but only because they are in the whole (*hoti en tôi holôi*). I take Aristotle to mean that they *contribute* to the perceptibility of the whole. He puts the point also by saying that our sight encounters them (*epelêluthen*, 446a1), and that they are potentially visible (446a5). Similarly in our *Physics* passage, Aristotle says that a small enough portion of millet might not, however long you gave it, move as much air as the bushel moves. Nor can one say that it moves as much air as it would when separated, for it could not exist separately, but exists only potentially within the bushel. Hence Zeno is wrong that it makes a sound.

Simplicius preserves a dialogue (1108,19-28) in which Zeno is presented as directing his paradox against Protagoras. He comments that Aristotle must be right about thresholds, or one could, given enough time, move Mount Athos a fractional distance, as Archimedes claimed; 'Give me somewhere to stand, and I will move the earth' (1109,32-1110,5). On the other hand, he says, Aristotle's proportionalities are meant to hold for situations above the threshold (1106,17-26).

Simplicius' Neoplatonism, as Hagen shows below, is apparent at several points. He claims that Aristotle's denial of self-motion applies to bodies, and does not exclude Plato's self-moving souls (1140,30-1041,1;1041,28-31). In his discussion of non-qualitative changes, he claims that a human soul that has lost its natural virtue is not a soul in the strict sense, but is dead (1066,21-7; cf. 1021,16-18). In the same context, discussing the acquisition of knowledge, he rejects Alexander's interpretation of Aristotle as saying at A 247b6-7 (it is still clearer in the B version at 247b20-1) that we acquire knowledge of the universal empirically through perception of particulars. Simplicius prefers to interpret Aristotle as taking Plato's view that the intellect knows particulars through universals which it knows non-intermittently (1074,6-1075,20). This is in spite of the fact that elsewhere Simplicius takes the view shared by almost all Neoplatonists that we have knowledge of universals through both routes. That is, we acquire universals empirically from perceived instances, although we may have to correct them by reference to the more accurate universals that we inherit from a previous life.[3]

Later in the same context, Simplicius takes up Aristotle's statement that the young adult's acquisition of knowledge is like recovering from a bodily illness, or drunkenness, or sleep, since the disturbances of childhood inhibit knowledge (A 247b16-248a3; B 247b28-248a3). Aristotle's aim is to deny that the acquisition of

3 Ian Mueller, 'Aristotle's doctrine of abstraction in the commentators' in Richard Sorabji, ed., *Aristotle Transformed: The Ancient Commentators and Their Influence*, London & Ithaca NY 1990, ch. 20. I have benefited also from unpublished work by Frans de Haas. The Simplicius references are Simplicius *in Cat.* 84,23-8; Simplicius (?) *in DA* 3 233,7-17; 277,1-6; 277,30-278,6.

knowledge is a qualitative change. Unlike Philoponus commenting on the same passage, Simplicius suggests that the argument will only work if the soul has always possessed knowledge, so that its learning is, as Plato held, recollection from before birth (1079,6-12). Although Simplicius is following the Neoplatonist policy of harmonising Aristotle with Plato, he lets it be known that there are other interpretations.

Translator's Note

The translation strives for faithfulness to what Simplicius says, without, it is hoped, unduly sacrificing readability. Important Greek words have been rendered in a uniform and consistent way, and the Greek-English Index and the English-Greek Glossary provide full details about how a given term has been translated. Simplicius' commentary already comes equipped with lemmata which indicate the portion of the text to be discussed. These have here been supplemented by a summary of the relevant section of Aristotle's text. These summaries have been inserted into the gaps in Simplicius' lemmata and are indicated by curly { } brackets. It may seem perverse that these summaries are occasionally as long as their Aristotelian originals, but *Physics* 7 is exceptionally difficult and some additional orientation and interpretation can be helpful to the reader of Simplicius' commentary. Square [ ] brackets enclose words which are not specifically in the Greek text but which seem necessary to a proper understanding of Simplicius' thought.

My greatest debt is to Richard Sorabji, who agreed to let me work on this project and who has been unfailingly patient and supportive. His capable research assistants, including Paul Opperman, Ian Crystal and Dirk Baltzly, have faithfully kept up a steady stream of correspondence. David Konstan, Robert Wardy, Catherine Atherton, and some very learned but anonymous reviewers have been kind enough to go over the translation and suggest manifold improvements, and I am most grateful to them. David Konstan's own excellent translation of Simplicius' commentary on *Physics* 6 has been a constant source of guidance and inspiration, as have his words of oft-repeated and friendly encouragement. In addition to this help with the translation, Robert Wardy has put all students of Aristotle in his debt by writing a splendid and illuminating book on *Physics* 7. Simplicius himself, ever the solicitous commentator and concerned about what was best for his readers, would no doubt have enthusiastically referred them to it. He would probably also have welcomed the Ibycus Scholarly Computer, and the modern who attempts to navigate the ocean of Simplicius' voluminous commentaries certainly has good reason to appreciate it. It has been invaluable on this project, and I am grateful to David Packard and the Ibycus Corporation for developing such a boon to scholarship. Similar thanks go to Professor Theodore Brunner and the staff at the Thesaurus Linguae Graecae for their wonderful CD ROM disk. Cal Poly granted me a sabbatical leave during which the project was undertaken. Much of that leave was spent in London, and while there I was able to use the joint library of the Hellenic and Roman Societies and the Institute of Classical Studies. Many thanks to the librarians and in particular to the Hellenic Society for all manner of help and hospitality. Finally, there is no way of expressing how much I owe to the love and patience of my wife Margaret.

The general editor wishes to thank Robert Wardy and Catherine Atherton for their detailed initial comments on the work in progress and the readers who commented on the final draft: Peter Lautner, Alan Lacey, Donald Russell and Christian Wildberg.

Textual Emendations

1038,23	reading *diorthôsas* instead of *diarthrôsas*.
1040,3	reading *AC* instead of *AB*.
1042,22	reading *kinoumenou* instead of *kinoumenon*.
1042,25	omitting <*to*> and reading *kinoumenou* instead of *kinoumenon*.
1046,18	reading *dunaton* instead of *adunaton*.
1048,17	reading *to hama* instead of *ta hama*.
1050,6	reading *auto* instead of *auton*.
1050,22	reading *hosai* instead of *hosa*.
1059,8	reading *nou* (as suggested by Diels) instead of *nun*.
1061,23	reading *alloioutai* instead of *legetai*.
1064,26	reading *alla mên oude* instead of *alla mên oute*.
1066,14	correcting *ekeisaktos* to *epeisaktos*.
1067,15-16	reading *kai* instead of *kai kai*.
1068,22	reading *allou* instead of *alloiou*.
1069,8-9	moving comma from after *metabolê* in line 9 to after *ên* in line 8.
1071,26	correcting *to* to *ta*.
1073,26	reading *oute* instead of *oude*.
1075,9	correcting *epistêmê* to *epistêmêi*.
1080,16	omitting <*tôn*>.
1080,22	reading *lambanein* instead of *analambanein*.
1083,19	reading *tou touto einai* instead of *toutou to einai*.
1084,16	reading *pote men meizôn* instead of *pote meizôn*.
1091,20	inserting closing parenthesis after *leuka*.
1091,21	inserting opening parenthesis before *hoion*.
1093,17	reading *anthupopherei* instead of *hupopherei*.
1094,17	reading *isos* instead of *isôs*.
1094,21	reading *hote* with Ross instead of *hoti*.
1095,28	reading *zêtômen* instead of *zêtôi men*.
1096,16	reading *hen tôi eidei* instead of *en tôi eidei*.
1097,20	reading *ei* instead of *hêi*.
1098,16	replacing the question mark with a comma.
1100,17	correcting *sumbekêkos* to *sumbebêkos*.
1101,1	reading *ei ison* instead of *hêi ison*.
1101,2	inserting smooth breathing mark in *tauta*.
1102,29	reading *houtos* instead of *autos*.

1103,32 reading *diastêma ti* instead of *diastêmati*.
1115,14 inserting <*hoion*> instead of <*kai*>.

Simplicius

On Aristotle Physics 7

Translation

The Commentary of Simplicius the Philosopher: Part Seven, on Book Eta of Aristotle's *Physics*

[CHAPTER 1]

This book of the *Physics*,[1] which it is the Peripatetics' custom to label 1036,3 Eta,[2] is the seventh, but it is transmitted in two versions, having some slight difference with respect to wording only,[3] for the same problems[4] and their demonstrations are transmitted in the same order in both. I too have now taken up the [material] from both which Aristotle's commentators clarify.

One ought to know that the more important and more germane to the present treatise of the problems found in this seventh book are found along with more exact demonstrations in the book after this one, the final [book] of the whole treatise.[5] This is why this book seemed to some to be redundant in [the context of] the treatise, inasmuch as it uses weaker or, as Alexander says, more logical[6] demonstrations; and Eudemus, at any rate, having up to this [stage] followed closely the main points of nearly the whole treatise, passed by this [book] as otiose and proceeded to the ones in the final book.[7] Themistius too, having arrived at this book in the course of paraphrasing the whole treatise, ignored many of the main points in it.[8]

But since this [book] too is plainly not at all lacking affinity with the other books in this treatise, nor unworthy of Aristotle's acumen, perhaps I might suggest that this [book] was written earlier by 1037,1 Aristotle, but, inasmuch as the main points in this one were later articulated more exactly in the final book, some included it in the treatise as germane.[9] It would also have, I think, a usefulness which is not to be dismissed, training us in advance and accustoming [us] to the consideration of the theorems Aristotle offers in the final book, which are genuinely great and hold together the whole doctrine of nature [*phusiologia*]. For example, the first [point] put forward by him in this [book, namely,] that everything moving is being moved by something,[10] has received a more exact demonstration there.[11] But as long as we are following closely the things in fact offered here, let us trace out their purport as much as possible.[12]

241b34-44[13] It is necessary that absolutely everything moving is being moved by something.[14] {If it does not have the source of motion in itself, it is evident that it is being moved by something else. On the other hand, if it does have the source of motion in itself, to suppose that it is being moved by itself and not by something else would be, first, like denying that a line KM, assumed to be moved by a part of itself KL, is being moved by something just because it is not evident which part of KM is the mover on account of its not being evident which is the mover} and which [is] what is being moved.

The point proposed is that everything moving is being moved by something,[15] on which all the subsequent theorems having to do with nature depend.[16] But since some moving things are being moved from outside, whereas others are not [being moved] from outside (those being moved by force [are being moved] from outside, while bodies moving in accordance with nature and those moving in accordance with psychic impulse[17] are not [being moved] from outside),[18] he did not deem it worthwhile to prove that things being moved from outside are being moved by something, inasmuch as what is moving [them] either by pushing or pulling or carrying or turning[19] is manifest.[20] On the other hand, he does prove it in the case of that which seems to be moving due to itself and not [being moved] from outside, having earlier distinguished that which is moving in its own right and primarily from things which are moving in virtue of another and, in general, incidentally, such as, for instance, things moving in virtue of a part, as they were distinguished at the beginning of the fifth book.[21] For, in fact, a person moving his hand is [not] himself said [to be moving in his own right and primarily,][22] but rather he is himself said to be moving because one of the things belonging to him is moving.

Having taken, then, a thing moving in its own right and primarily due to itself, which is moving neither incidentally nor because one of the things belonging to it is moving, he proves that in what is moving in this way the mover is one thing and what is being moved is another.

First he guides our thinking aright, teaching [us] not to believe that a thing which is moving as a whole due to itself is not being moved by anything [just] because the thing moving it is not apparent
1038,1 from outside. For it is possible for the mover not to be evident even in cases in which the mover is obviously one thing and what is being moved by it another, as he himself shows in the case of an illustrative example.[23] For if KM should be moving as a whole because part of it, KL, which is itself moving too, is imparting motion, whereas the remaining part of it, LM, is being moved by KL, one could not justly

claim that KM is not 'being moved by something on account of its not being evident which [thing]' in it 'is the mover and which is what is being moved' [241b42-4]. For, in fact, in the case of animals, the animal is seen moving as a whole, but it is agreed that the body is moved by the soul (which, because the body is moving, is moving along with it).[24] This is why, even if [the soul] is not apparent, we say that the body is moved by it. Thus, then, when AB too is moving as a whole, even if the mover is not apparent, one must not for this reason claim that it is not being moved by anything.

If, then, a thing moved from outside is obviously moved by something, and also nothing prevents a thing [moving] due to itself from being moved by something, even if the mover is not apparent, it is inferred that everything moving is being moved by something, since the mover's not being apparent is no hindrance to the argument [*logos*] for us.

It is obvious that it has not been proved on the basis of this argument [*epikheirêma*] that a thing which seems to be moving due to itself is also of necessity being moved by something, but we do know that, even if the mover is not apparent, nothing prevents what is moving's being moved by something. The next [passage] proves the point proposed as follows:

> **241b44-242a49**[25] Then, [it is not necessary that] a thing which is not being moved by something {ceases moving because another thing is at rest, and hence anything which is at rest because another thing has ceased moving must be moved by something. If this is assumed, then everything moving is being moved by something. If AB is moving, it will be divisible, since everything moving is divisible. Let it be divided at C. If CB is not moving, then AB will also not be moving, because if AB were moving then AC would be moving while CB was at rest and hence AB would not be moving in its own right and primarily, as hypothesized. It was agreed that anything which is at rest because another thing is not moving is moved by something. Consequently, it is necessary that everything moving is being moved by something, for that which is moving will always be divisible and if a part of it is not in motion} it is necessary for the whole too to be at rest.

Having proved that it is not necessary, if the mover is not apparent, that what is moving is not being moved by anything, and having corrected[26] the preconception deriving from the appearance, he next proves the point proposed, [namely,] that everything moving is being moved by something. He proves it after having assumed in advance that 'if something is at rest because another thing has ceased moving, it is necessary that it is being moved by something' [242a36-7]. For 'it is not necessary' that a thing which is not being moved by something 'ceases moving' [242a35] when some other thing has ceased. For though due to chance a thing which is not being moved by something might on occasion itself cease moving too when that [other thing] has ceased from motion, it does not, in fact, do so of necessity. Accordingly, if 'it is not necessary that a thing which is not being moved by something ceases [moving]' [241b44-242a35] when some other thing has ceased, 'it is necessary' that a thing which does of necessity cease [moving] when some other thing has ceased 'is being moved by something'.[27]

1039,1 Having inferred this, then, in accordance with the second [mode] of hypothetical [arguments],[28] and having assumed in addition that everything moving is divisible, which indeed he proved toward the end of the preceding book,[29] he sets out as an illustration[30] the moving thing AB, which is divisible because it is moving. He divides it at C and claims that if CB, which is part of ACB, is not moving, then neither will AB still be moving as a whole. For if anyone should claim that AB is moving though CB is not moving but only AC is, then AB as a whole will no longer be moving in its own right, and not primarily either, but rather because a portion of it, AC, is moving. But AB was not hypothesized to be moving in this way, but rather in its own right and primarily. Therefore, if CB is not moving, AB as a whole will not be moving primarily. But if AB is not still moving when CB is at rest, it will be being moved by something, for it was posited that a thing which ceases moving because something has come to a halt is being moved by another.[31]

The most scholarly[32] Galen[33] criticizes this demonstration, and others criticize it too, on the grounds that it uses an impossible hypothesis, [namely,] the one claiming that, though AB is moving primarily and in its own right, CB is not moving. On the other hand, even though suspecting the demonstration in many places, Alexander[34] chose to claim that it is not impossible for the purposes of hypothesis to hypothesize that part of a thing moving in its own right and primarily comes to a halt. 'Because', he says, 'for the purposes of hypothesis only things destructive of one another are impossible, as, for example, sailing through rock'.[35]

But [we may reply that] perhaps this too is [just] such a case: for if this thing moving in its own right and primarily is a thing itself moving as a whole and not one moving because some part of it is moving, how can [the fact that] one part of it is moving and another part is not moving coexist with [its being] a thing moving in its own right and primarily? But the entailment is valid, for if some portion of a thing moving in its own right should come to a halt, the whole too will come to a halt, and it as a whole will no longer be moving in its own right and primarily. It is impossible for some portion of a thing moving in its own right to have come to a halt as long as the thing is hypothesized to be moving in its own right and as a whole.

This too seems puzzling to me: In what way has ACB been assumed to be moving? For if it [has been assumed to be moving] only as a body, what in it would be the mover? But if [it has been assumed to be moving] as an animal, with CB moving only, on the grounds that it is a body, and with AC assigned to the animal's soul and for this reason simultaneously imparting motion and moving, how is it claimed that the whole is moving in its own right and primarily, since AC,[36] that is, the soul, is moving incidentally because it is in the moving body?[37] How could AC be moving at all if CB is at rest, since the soul is unmoved in its own right but moves because it is in a moving thing, [namely,] the body?[38] Consequently, AB could not even be claimed to be moving in virtue of a part on the grounds that AC in it is moving, for if CB (assigned to the body) is at rest, then AC (employed for the soul) would be at rest too.

They also find fault with the argument on the grounds that it begs the question, since it is through the part's not moving that it proves that the whole is not moving in its own right, for the [argument] hypothesizing that the part is not moving is thereby hypothesizing that the whole is also not moving in its own right and primarily.

Since it does not seem to me that any of Aristotle's thoughts is to be dismissed, and Alexander, even though he has twisted and turned in many ways, has plainly not articulated the argument as flawless, perhaps nothing prevents putting my own conjecture concerning it before my readers for examination as well.[39]

Maybe, then, Aristotle, having proposed to prove that that which is moving in its own right and primarily, being the body, does not have its source of being[40] from itself but is moved by something, proves this after having assumed in advance that anything which is at rest because some 'other thing has ceased moving' is being moved 'by something' [242a36-7] and does not have the motion due to itself and also does not move itself.

He proved this through its opposite, which is more evident, after having performed contraposition.[41] For if 'it is not necessary that a thing which is not being moved by something ceases[42] moving' because some 'other thing is at rest' [241b44-242a35], then that which does necessarily cease moving because some other thing is at rest is being moved by something.[43] As the other thing upon whose ceasing [to move] the whole that is moving in its own right and primarily ceases [to move], he took the part. This is why at the end of the argument he added: 'When the part is not moving, it is necessary that the whole too is at rest' [242a48-9].

Accordingly, he did not take the thing upon whose ceasing [to move] the whole ceases moving as the mover, but rather as itself also moving in the whole and completing the motion of the whole with its own motion. Yet if the whole were moving due to itself and not [being moved] by some other thing, but were itself both the mover and what is being moved, it would be necessary for it as a whole to coincide with itself as a whole and to be partless and indivisible,[44] and also, I will
1041,1 add, always to be moving: for inasmuch as it does not 'depart from itself it never leaves off moving'.[45] But if it is divisible, then, since it does not coincide as a whole with itself as a whole, the mover and what is being moved are not everywhere present to one another, but rather are nowhere present in the same [place]. This is why it is possible in the case of this to assume that some part is not moving and for the whole too to be at rest on account of the part.

Accordingly, [Aristotle] himself reasonably stated that the moving thing's being divisible is responsible for the whole's being at rest when the part is not moving, so that the inference involved in the argument, according to Alexander, is in the first figure as follows: 'Everything moving in its own right and primarily is of necessity not moving when something has ceased from motion. That which is of necessity not moving when something has ceased from motion is being moved by something.'[46]

But perhaps it is better to make the inference in the second figure[47] thus: 'Anything moving in its own right and primarily, being divisible, is at rest as a whole when a part is at rest. Anything moving due to itself is not at rest as a whole when a part is at rest (for a thing like that does not have a part).[48] Therefore, anything moving in its own right and primarily, being divisible, is not moving due to itself.'[49] The affirmative 'is being moved by something' [242a47] substituted for this [sc. 'not moving due to itself'] made Alexander infer the conclusion in the first figure. I judge that it has been inferred just as I have stated from [Aristotle's] having put the whole minor premise toward the end of the argument, when he says, 'For anything moving will always be divisible, but when a part is not moving it is necessary that the whole too is at rest' [242a47-9], which is the same, I think, as 'Anything moving being divisible, when a part is at rest the whole is at rest'.[50]

If in any way whatever I have hit upon the aim of what has been stated [by Aristotle], it is not impossible, I think, to hypothesize that a part of a thing moving in its own right and primarily is not moving.[51] For a thing which was divisible but had not yet been divided was hypothesized to be moving in its own right and primarily, but hypothesizing that a divisible thing has been divided is not among the things which are impossible. And hence Aristotle too, having hypothesized that a thing moving in its own right is divisible, divided it, and it followed from the actual division that the thing was not moving in its own right.[52]

Also, in reply to my puzzle,[53] I claim that the moving thing AB is a body,[54] which indeed he proved was not moving due to itself on the grounds of its being partitioned and the whole's being at rest when it has been hypothesized that a part in it is at rest. That does not pertain to anything moving due to itself on account of its not even having a part.[55]

Taken in this way, the argument could not be claimed to beg the question either.[56] For AB was assumed, as long as it was divisible and not divided, to be moving in its own right and primarily. But 1042,1
when the divisible thing was divided, the whole was proved to be at rest through the [state of] rest of the part, so that it was not still moving in its own right and primarily after actually having been divided. For either, if the parts which have actually come to be are moving, the whole is no longer moving primarily, or, if any of the parts is at rest, the whole is no longer moving in its own right, but rather, if at all, because some part of it is moving.

One ought to know, however, that in the next book[57] he again proposes to prove this [point] that everything moving is being moved by something, and he proves it more exactly and more effectively.[58]

242a49-b53[59] Since it is necessary that everything moving is being moved by something,[60] {if one thing is locally moved by another which is moving, and that in turn by another which is moving, and so on, it is necessary that there is a first mover rather than proceeding to infinity. For let there not be a first mover, and let the series be infinite. That is, let A be moved by B, B by C, C by D, and so on. Since it is hypothesized that each mover is imparting motion inasmuch as it is moving, it is necessary that the motions occur simultaneously. Hence, the motion of A, B, C, and each of the things imparting motion and moving will be simultaneous. Let the motions of A, B, C, and D be designated E, F, G, and H, respectively, for one can assume that there is a motion which is one in number belonging to each thing. Let K be the time in which the motion of A occurs; inasmuch as the motion of A is finite, K will also be finite. Since there are infinitely many movers and moving things, the motion EFGH made up of the motions of all the members of the series will be infinite.[61] The motions may be equal or some may be greater than others, but both ways the whole motion will be infinite. Since all the motions occur simultaneously, this infinite motion will occur in the same time that the motion of A does. But the motion of A occurs in a finite time, and consequently there will be an infinite motion in a finite time,} but this is impossible.

Having assumed as proved that everything moving is being moved by something, with this posited he proves that there is some first mover and that one thing does not move another [and so on] to infinity, but rather there is something which imparts motion though not being moved by another. He proves it in the case of motion with respect to place, for since this is the first among motions (because if this does not exist no other motion exists either, as we shall learn in the next book),[62] if it is proved in the case of this [motion] that there is a first mover, it would have been proved universally as well.

The proof proceeds through reduction to impossibility. For, having hypothesized that one thing is being moved by another [and so on] to infinity, he proves that it is consequent upon this that an infinite motion occurs in a finite time, which indeed has been proved impossible in the book before this one.[63] It is rightly stated 'by another which is moving',[64] lest by saying only 'by another' he seems to assume what [was to be proved] in the beginning, [namely,] that there is some first mover.[65] He proves it by having proved that it is not the case that there is one moving thing imparting motion before another [and so on] to infinity. He produces the proof, then, in the case of this [situation] of always being moved 'by another which is moving',[66] denying the 'to infinity' attaching to this and proving that there is some first mover which is not itself being moved by another.

Having hypothesized, then, that anything moving is being moved by another which is always itself too being moved by another, and this to infinity, he shows the impossibility consequent upon the hypothesis. For since the mover is hypothesized to be imparting
1043,1 motion inasmuch as it itself is being moved by something,[67] it is obvious that it will simultaneously be being moved by the thing moving it and moving the thing being moved by it, since it is imparting motion because of being moved by another. Therefore, the [motion] of the mover and the motion of what is being moved are accomplished in the same time and in the same respect (for it is when it is being moved that the mover is imparting motion).

Consequently, the motion of all the things imparting motion and moving will be simultaneous. For if A should be moved by B, and B by C, and the latter by D, 'and the next always by the next', it is evident that the motion of A and that of B and that of C and that of each of the things imparting motion and moving will be simultaneous, since each of the things imparting motion and moving is numerically one,[68] and the motion of a thing which is numerically one is numerically one. For even if the [motion] made up of all [the motions] comes to be one and continuous, still this single [motion] is composed of numerically as many [motions] as there are things imparting motion and moving. For even if all are moving simultaneously, still each of them is moving [with its] own motion, inasmuch as they are numerically distinct, for the motion of a thing which is one in number is numerically one and determined by the single thing.[69] 'For every motion', he says, 'is from something to something' [242a65-6]. For the essence of motion and of change in general is being 'from something to something' and being determined by that from which and that to which.

Having stated that the motion of each of the things imparting motion and moving is one in number, he points out that the [motion] occurring 'from the same thing to' numerically 'the same thing in' numerically 'the same time' [is one in number].[70] By 'from the same thing' he means that from which, and by 'to the same thing' that to which, whether that from which and to which are the same or different. Also, if the time of the motion from something to something is not one and continuous but leaves a gap, the motion is not one. For if a thing does not change 'from this white to this black' [242b39] in a time which is numerically one and continuous, but rather in different [times], then, even if that from which and that to which are as numerically one as possible, and the changing thing is one too, the interrupted motion will not be numerically one, but rather, if [it is one] at all, one in species.

Having spoken about motion which is numerically one, he adds that motion too, like other things, has unity in three ways, for [it has unity] either in number, as has been stated, or in genus or in species. It has been stated what [motion] is numerically one, whereas motion which is one in genus is that 'belonging to the same category, for instance, substance' [242b35], like coming to be and perishing,[71] 'or quality' [242b35], like alteration, or quantity, like increase and decrease, or the [category of] where, like [motion] with respect to place (for it is his custom to speak of these genera as 'categories').[72]

[Motions] occurring from things which are the same in species to things which are again the same with one another in species are motions which are the same in species, for instance, those occurring from white to black, for the white is of the same species and the black is of the same.[73] Accordingly, whether several motions from white to black should occur simultaneously, or whether [they should occur] during different times, they are the same in species.

Having said 'or from good to bad', he added 'if it is undifferentiated in species',[74] obviously meaning good and bad in the sense of, for example, temperance and intemperance. If, however, the goods differ with respect to species, the ones from which and again the ones to which the change [occurs], as, for example, temperance [differs] from courage and intemperance from cowardice, the changes in such things are no longer of the same species, but rather, if [they are the same] at all, of the same genus in virtue of the good and bad.

I happen to have spoken before concerning motion which is numerically one,[75] and one must add only this much: the moving thing too must be numerically one if the motion is going to be numerically one. The unity of motion[76] has also been discussed in the fifth [book] of this treatise, to which he referred us when he stated that it has been discussed 'in [what was said] earlier' [242b42].[77]

After establishing that each of the things imparting motion and moving is moving [with] a motion which is numerically one, he takes the time in which one of the moving things (for instance, A) moves, which is finite, since the motion of each is also hypothesized to be finite, inasmuch as it is 'from something to something' [242a66]. It has been proved that a finite motion [occurs] in a finite time.[78] He posits that the time K of the motion of A is finite. Since the things imparting motion and moving are infinite, [it follows that] even if the motion of each is finite, nevertheless 'the [motion] made up of absolutely all [of them]', inasmuch as they are infinite, 'will be infinite' [242b46-7]. For it is necessary that the combined magnitude made from the combining of infinitely many magnitudes is infinite, even if each of the things which are combined is finite. Accordingly, the motion EFGH, to state it in a word,[79] constructed from infinitely many motions, is infinite. Since, then, in the time in which A moves, that is, in K, each of the other moving things, which are infinitely many, moves too, it will result that the whole motion EFGH, which is infinite, also occurs in the finite time K.

Therefore, something impossible did follow, [namely,] moving [with] an infinite motion in a finite time, for this was proved impossible in the book before this one.[80] It followed from hypothesizing that the things imparting motion and moving are infinitely many. Therefore, it is impossible that these things are infinite. Therefore, there is some first thing which imparts motion though not being moved by another.

Since what is made up of infinitely many magnitudes is infinite, whether the magnitudes are equal or unequal, and motion too is a kind of magnitude, he reasonably added that 'whether' the infinitely many motions are 'equal' to one another 'or' some are 'greater' than others, 'both ways the whole is infinite' [242b49].

242b53-9[81] The original [point proposed], then, might seem to have been proved in this way, {but in fact no impossibility has been shown. It is possible for there to be an infinite motion in a finite time, provided it belongs not to a single thing but to many, which is just what has resulted. Each thing is moving with its own motion,} but it is not impossible for many things to be moving simultaneously.

Inasmuch as it was assumed in an indefinite way in the demonstration just stated that the things imparting motion and moving are infinitely many, he rightly remarks that without receiving appropriate additional specification the argument [only] seems to demonstrate what was originally proposed, [namely,] that there is a first mover, by proving that an impossibility is consequent upon the opposite claiming that before each thing there is always another which is both imparting motion and moving [and so on] to infinity and for this reason hypothesizing that the motions are infinite and constructing a single [motion] infinite in magnitude, [namely,] the one made up of the infinitely many [motions], and inferring an impossibility, [namely,] that this infinite [motion] occurs in a finite time.

Accordingly, he proves that an impossibility is not inferred from these posits. For what was posited was that the motions of infinitely many things were themselves infinitely many too, but it is possible for these infinitely many motions to occur in a finite time if they occur simultaneously, as is stated, [i.e.] the [motion] of A and B and the others. And nothing impossible follows, but rather the impossibility followed from moving [with] a single infinite motion in a finite time, since, at all events, it is not impossible for several motions, and even infinitely many, to occur simultaneously in a finite [time].

If, then, the impossibility is going to follow, one must prove how the infinitely many motions are combined into a single motion infinite in magnitude occurring simultaneously in the finite time in which one of the finite [motions] also occurs, for thus the impossibility that there is an infinite motion in a finite time would follow. Accordingly, 1046,1
he does this next, showing how such a motion is rendered a single infinite one.

242b59-243a31[82] But if [it is necessary that] the thing first[83] imparting motion with respect to place {must be either touching or continuous with what is being moved, as we see in every case, then there is a single thing made up of all the things which are imparting motion and moving. (It makes no difference for present purposes whether this thing is finite or infinite, for the motion of infinitely many things will invariably be infinite, since it is possible for them to be both equal to and greater than one another.) We will assume that what is possible is actually the case. If the thing made up of A, B, C, and D moves with the motion EFGH in the finite time K, it results that either a finite or an infinite thing goes through an infinite motion in a finite time, and it is impossible both ways. Consequently, it is necessary for there to be some first thing imparting motion and moving. It makes no difference that the impossibility has resulted from a hypothesis. The hypothesis has been assumed possible, and when something possible is posited} nothing impossible should occur on account of it.

Though the things imparting motion and moving are many and infinite in multitude, nothing impossible is consequent upon the infinite motions' occurring simultaneously in a finite time, but a single thing does come to be, he asserts, [namely,] the one made up of all the things imparting motion and moving, for it is necessary that the thing first moving something with respect to place (that is, proximately and not by means of another) and imparting 'bodily motion' [242b60] moves it by being continuous with what is being moved or by touching it. By 'bodily motion' he means that involving resistance, for the first mover and the beloved also impart motion,[84] and they impart motion with respect to place, but not in a bodily way, for it is what is pushing or pulling or turning or carrying[85] that imparts motion in a bodily way. It is necessary, then, that such things touch the things being moved by them and through the contact be united with them in some manner. Accordingly, whether the things imparting motion and moving in succession are several or whether [they are] infinite, they too will be one in virtue of contact.

'It makes no difference', he says, 'whether this' single thing is 'infinite or whether [it is] finite',[86] not because it is possible[87] for some single finite thing to come to be from infinitely many magnitudes touching one another, even if they should be smaller in magnitude than Democritus' atoms, but because it is not proposed to consider here whether any infinite magnitude does exist or whether it does not exist. For even if someone hypothesizes that it is finite, [nevertheless,] inasmuch as the motion composed of the infinitely many motions is infinite and one, the impossibility will follow, [namely,] having moved [with] an infinite motion 'in a finite time' [242b69-70], whether 'the finite' magnitude or whether 'the infinite' one [has done so], for 'it is impossible both ways' [242b70-1], as has been proved in the book before this one.[88]

He mentioned here too that no matter how the infinitely many motions are taken, whether as being equal or whether [as being] unequal with respect to magnitude, they make the whole infinite in
1047,1 magnitude. He added here that in fact the motion becomes one on account of the touching of the things imparting motion and moving, which makes the moving thing one. But the motion of a single thing is one.

He stated, 'For we will take what is possible as being the case' [242b66-7], concerning taking [the motions] to be equal or unequal. For since it is possible both ways, whichever is hypothesized we will have the impossibility following from it. Having hypothesized that the magnitude made up of ABCD is infinite[89] (there being infinitely many magnitudes), in the conclusion of the argument he said 'either the finite or the infinite' thing [242b70-1], on the grounds that the impossibility follows with respect to each. For, in fact, in the preceding book it was proved equally impossible for an infinite and a finite magnitude to move [with] an infinite motion in a finite time.[90]

After resolving the objection based on what was hypothesized in an indefinite way, he next draws the conclusion of the argument. For, inasmuch as it has been proved impossible for the things imparting motion and moving [to go] to infinity, since an impossibility followed from it, it is necessary that the opposite of this be true, [namely,] that this procession comes to a halt and does not go to infinity, but rather there exists something first imparting motion which is no longer being moved itself by another.[91]

Since his argument did not proceed directly [*deiktikôs*],[92] but rather on the basis of a hypothesis that the things imparting motion and moving are infinitely many, and inferred[93] the impossibility with that, in order that no one might state that nothing has been proved, since the impossibility followed from a hypothesis, he rightly added that 'the hypothesis has been assumed as a possibility'.[94] For the person who does not admit that there is some first mover, in the belief that it is possible to assume that [the series of] things imparting motion and moving, one before another, is able [to go] to infinity, does not thus admit [a first mover], but when what is possible has been hypothesized nothing impossible follows; rather, the impossible [follows from] the impossible, and the possible from the possible, as we have been taught in the *Analytics*.[95] Consequently, even if it was hypothesized as a possibility, it was not possible.[96]

When [Aristotle] says, 'For the hypothesis has been assumed possible' [243a30], one must understand 'as a possibility' instead.[97] Consequently, even if the impossibility resulted from a hypothesis, it resulted from a hypothesis which was hypothesized as a possibility. And if indeed it were possible, something impossible should not have followed from it.

[CHAPTER 2]

243a32-40[98] The first mover, not in the sense of that for the sake of which {but in that of the source of motion (i.e. the efficient cause) is together with what is moved, and there is nothing in between them. This is common to everything being moved and imparting motion. Since there are three kinds of motion, that with respect to place, that with respect to quality, and that with respect to quantity, there are three kinds of movers. Let us speak first concerning locomotion,} for this is the primary motion.[99]

1048,1 In order to prove that the motion based on the hypothesis that the things imparting motion and moving [extend] to infinity becomes one in number and with this to infer an impossibility, [namely,] moving [with] an infinite motion in a finite time, he used the proximate mover's being obliged to touch the thing being moved or to be continuous with it.[100] He confirmed this then by induction, saying 'just as we see in all cases' [242b61]. But now he proposes to demonstrate with respect to each species of motion that it is necessary for the proximate mover to be together with what is moved.[101]

Since movers are of two kinds,[102] on the one hand as the source of motion (that is, the efficient [cause]) and on the other hand as that for the sake of which (that is, the final [cause]), his demonstration here concerns the efficient [cause]. For that which imparts motion as an end and, speaking generally, the desired are not together with what is moved, for desire is above all for what is not present, being a stretching out toward it.[103]

The thing supplying the cause of the motion, even if by means of some other motions in between, is also spoken of as a first mover, as for example the person who is pulling by means of ropes or pushing by means of a rod. In this way too, the very first, unmoved cause of motion is also spoken of as moving the things here[104] [in this world] as an efficient [cause]. The proximate mover too is spoken of as a 'first' mover in the sense in which he now used the term when he said, 'The first thing imparting motion' in this way is 'together with what is moved' [243a32-3].

He himself explained how he means 'together' [*hama*][105] when he stated, 'I mean by together that nothing is in between' [243a33-4], which is the same as touching, for those things were spoken of as touching 'whose extremities are together' so as to have nothing in between.[106]

After stating, 'This is common to everything being moved and imparting motion'[107] (imparting motion in this way, obviously, as has been stated, in the manner of an efficient cause), he next divides the species of motion into that with respect to place, with respect to quality, and with respect to quantity. Having accordingly taken the movers to be three, [namely,] 'the thing causing locomotion, the thing causing alteration, and the thing causing increase or diminution' [243a38-9], he produces the argument first in the case of locomotion, on the grounds that it is the primary motion, as he will prove in the next book.[108]

243a11-b10[109] Absolutely everything moving locally, then, is moved either itself[110] by itself {or by another. In the former case it is evident that the thing being moved and the mover are together. In the latter case, locomotion can occur in four ways: pulling, pushing, carrying, or turning. All motions with respect to place are reduced to these. Pushing on, pushing away, and throwing are pushing. Pushing apart is pushing away, and pushing together is pulling. Consequently, the species of pushing apart and pushing together are also reducible in this way, for instance, tamping the woof and parting the warp. Similarly for all other instances of aggregating and separating, since all of them will be either pushing apart or pushing together,} except those [involved] in coming to be and perishing. 1049,1

By 'moving locally' [*pheromenon*] he means that which is moving with respect to place and is so designated from locomotion [*phora*]. It is obvious that if there is a thing which is moving it is necessary that there is also a thing which is imparting motion, since it has been proved that everything moving is moved by something.[111]

Having divided things moving with respect to place into those moved by themselves and having the mover in themselves (as animals have in themselves the soul moving the body) and into those moved from outside and by another (for it is not possible to be moving besides these ways),[112] he asserts that the things moved by themselves evidently have the mover together in themselves with what is moved, since [the mover] is neither separated nor divided from it by anything inserted in between. For the soul is associated with the body in this way when moving it, even if [the soul] is not imparting motion in a bodily way.

On the other hand, motion produced by something imparting motion from outside, being forced, obviously, and not in accordance with nature, has four highest varieties [i.e. summa genera]: for the mover is either pulling or pushing or carrying or turning [something]. He proves that these are the only types of forced motion produced from outside and by another on the basis that all motions produced by another by force are classified under[113] one of these varieties.

For he classifies pushing along [*epôsis*] and pushing away [*apôsis*] under pushing [*ôsis*]. For if pushing, as he himself defines it a little later [244a7-8], is either from a thing itself toward another or from one thing toward another, and there is pushing along whenever a thing, as well as following something moving away from itself, is pushing it along, and there is pushing away whenever something is being pushed either away from [the mover] itself or away from another thing by a mover that is not following after the moving thing, then pushing is common to both, but pushing along and pushing away differ because of the following after or not.

He asserts that throwing too is reduced to pushing: for something is said to be thrown as long as the vehemence of the pushing prevails over the proper motion of the thing thrown. But when that [vehemence] has been exhausted and this [proper motion] prevails, the stone, for instance, is no longer borne sideways, perchance, or upward due to the throwing, but downward due to its own proper inclination. Accordingly, throwing too is a kind of pushing, for even if someone throws the stone downward, the motion is throwing as long as the pushing of the thrower prevails over the natural motion. But when 1050,1 that [pushing] has been exhausted and the natural [motion] dominates, it is no longer throwing, and not pushing either.

Also, in the case of throwing, even if the thrower is no longer touching what has been thrown, nevertheless the air proximately moving it, at all events, having taken [its] source from the first thing which imparted motion and being naturally well-suited to motion to the side [sc. horizontal motion] on account of being neither absolutely heavy nor light,[114] pushes what has been thrown since it is touching it, just like the human being, perchance, who first moved it [sc. what has been thrown][115] or the air which received the power of imparting motion in succession from that person owing to continuity. For in the next book [Aristotle] himself accounts for throwing in this way, not accepting reciprocal replacement [*antiperistasis*].[116]

He also subsumes pushing apart and pushing together under pushing and pulling. 'For pushing apart' is 'pushing away' [243b3-4] from one another, and pushing away is pushing, whereas 'pushing together is pulling', and pulling is either 'toward itself' or 'toward another' [243b5-6]. 'Pushing together' [*sunôsis*] also seems to indicate a kind of pushing [*ôsis*], but not a separative kind like pushing apart but rather one which brings together and which tends to compress.

The species of pushing apart and pushing together, he asserts, will also be classified under pushing and pulling. Parting the warp and spitting and exhaling and every segregative motion are species of pushing apart, while tamping the woof and inhaling and drawing in food or drink and, simply, every aggregative motion are species of pushing together. Lest, by going over the individual [species], he prolong the argument and at the same time furnish suspicion about those left out, having reduced all such [motions] to aggregating and segregating he asserts that segregating is a kind of pushing whereas aggregating is [a kind of] pulling.

Having stated that all aggregatings and segregatings will be pushings together and pushings apart, he added 'except those[117] which are [involved] in coming to be and perishing' [243b9-10]. [He says this] since[118] it seemed to the circles of Anaxagoras and Leucip-

pus and Democritus[119] that all coming to be and perishing occurred by aggregating and segregating.[120] This is not his view,[121] but if, he asserts, any things at all do come to be and perish in virtue of aggregating and segregating, this aggregating and segregating could not be reduced to pushing together and pushing apart, nor through them to pulling and pushing. For what is being pushed together and pushed apart are spoken of as pushed apart and pushed together and in general as pushed and pulled while existing in accordance with [their] substance and remaining just what they are, whereas things that are coming to be and perishing, which do not remain in their own 1051,1
proper substance, change in virtue of coming to be and perishing, since such changes are with respect to substance. This is why such aggregatings and segregatings are not pullings or pushings, for comings to be and perishings are not even motions at all,[122] nor can what is still coming to be and not yet existing be pulled or pushed.

Since Alexander claims that in the other seventh book, which differs slightly from this one with respect to wording,[123] both tamping the woof and parting the warp are subsumed under pushing, one ought to know that in the [copies] I am acquainted with I have found it written thus: 'and both tamping the woof and parting the warp, for one of them is aggregating, the other segregating.'[124]

One ought to know that Themistius, in the books which I am acquainted with, began to paraphrase this book [starting] from this passage that says, 'Absolutely everything moving locally, then, is being moved either itself[125] by itself or by another' [243a11-12], having disdained the things stated in it up to this [point], and he also does not maintain the continuity in what follows.[126]

> **243b10-16**[127] At the same time it is evident that [aggregating and segregating] are not some other genus [of motion] {different from these four, because all instances of aggregating and segregating fall under the four kinds stated. Inhaling is pulling, and exhaling is pushing, and similarly spitting and other eliminative or assimilative motions of the body, which are instances of pulling on the one hand and pushing away on the other. One must classify the other motions with respect to place in this fashion,} for absolutely all fall under these four.

He has stated that aggregatings and segregatings are pushings together on the one hand and pushings apart on the other, and hence pullings on the one hand and pushings on the other. Since some of the physicists[128] posited aggregating and segregating as the principles not only of motions with respect to place but also of all [motions], and not only of motions but of all changes as well,[129] he himself seems to deny not only that they are the principles of all motions, but, he asserts, they are not even first principles, nor is this at all some genus of motion besides the four stated, but these [sc. aggregating and segregating] too and their species are classified under those. For, in fact, inhaling, which is aggregating, is pulling, whereas exhaling, which is segregating, is pushing. 'Similarly, spitting too and', he says, 'those other motions through the body' which are 'eliminative or assimilative' [243b13-14], the former being segregative and the latter aggregative, are classified under pushing on the one hand and pulling on the other.

Alexander, then, explained [the passage] in this way to begin with,
1052,1 claiming that aggregating and segregating have been proved not to be another genus of motion but themselves to be subsumed under the aforementioned four as parts of those.[130] But next, having taken up the argument again, whether owing to another view appended later or just how [it came about] I do not know, he asserts that aggregating and segregating are proved by [Aristotle] to be the most generic motions. 'For', [Alexander] says, 'all the species of motion will be found under these, for if all [come] under pushing apart and pushing together, and of those one is aggregating and the other is segregating, all would be reduced to aggregating and segregating.'

Alexander wrote these things [using this] very wording, even though Aristotle stated, 'Similarly too the other instances of aggregating and segregating, for absolutely all will be pushings apart and[131] pushings together, except those [involved] in coming to be and perishing' [243b7-10]. Consequently, pushing together is not a kind of aggregating, nor is pushing apart a kind of segregating, as Alexander asserts, but, on the contrary, aggregatings and segregatings are pushings together and pushings apart.

How could all motions be reduced to aggregating and segregating on the basis of what has been stated here, since Aristotle asserts that all [motions] fall under the four stated and, having in turn reduced two of the four, carrying and turning, to the remaining two, pulling and pushing, formulated the present argument for these [viewed] as general [kinds], stating, 'Consequently, if what is pushing and what is pulling are together with what is being pushed and pulled, it is evident that there is nothing in between the thing being moved with respect to place and the mover' [244a4-6]. He did not state, 'Consequently, if what is aggregating and segregating ...'.

However, in the other seventh book it is written with the following wording: 'And every motion with respect to place, then, is aggregating and segregating' [B 243b29]. Perhaps based on this wording Alexander added what was stated later, though that, as it seems, is not in harmony with what has been written here.[132]

Perhaps it is possible to prove that the accounts are in harmony,

if Aristotle is not simply classifying aggregating and segregating under pushing together and pushing apart, or rather under pulling and pushing, as under more general [kinds], but claiming that these [i.e. pulling and pushing] are the same as those [i.e. aggregating and segregating]. For Alexander too put it this way later when he said, 'All [motions] would be subsumed under these two, [namely,] pulling and pushing, of which the former is aggregating and the latter segregating. Thus aggregating and segregating would be the most 1053,1
generic motions.'[133]

243b16-244a11[134] Of these, in turn, carrying and turning {fall under pulling and pushing. Carrying occurs in virtue of one of the other three types of motion, since what is being carried is moved incidentally because it is in or on something being pulled or pushed or turned. Turning is composed of pulling and pushing, for the thing causing the turning is pulling one part toward itself and pushing another away from itself. Consequently, if that which is pushing and that which is pulling are together with what is being pushed and pulled, respectively, it is evident that nothing is in between what is being moved with respect to place and what is imparting motion. This is also obvious from the definitions. Pushing is motion either from a thing itself (or from another thing) toward another. Pulling is motion from another toward itself or toward another, when the motion is faster than the motion separating the things,} for thus the other thing is pulled along after [the mover].

Having proved that all motions with respect to place are reduced to the four stated, [Aristotle] in turn reduces two [of those], carrying and turning, to the remaining two, pushing and pulling. For he proves on the one hand that carrying occurs in virtue of pulling or in virtue of pushing or in virtue of turning, and on the other that turning is composed of pulling and pushing.

He proves that carrying [comes] under one of the three by assuming that what is being carried is moving not in its own right but incidentally, for it is because the carrier is moving that what is being carried is moving, and as the former is moving in its own right so is the latter incidentally. Consequently, one must inquire how many ways carriers move, for what is being carried will also move in the same ones.

Some carriers carry while moving due to themselves, as a horse [carries] a rider and in general animate things do, whereas others are themselves too being moved from outside. But concerning animate things and things moving due to themselves it has been proved earlier that they have both the mover and the moved together in themselves [cf. 243a12-15]. On the other hand, things which are inanimate and moved from outside, among which some are carriers, move in the three ways, for they are either pulled, as, for example, a ship carrying passengers; or pushed, as a [ship] sailing before the wind; or being turned, as tops and millstones which also sometimes carry things lying on them. Consequently, carrying is common to the three ways [in which things move with respect to place], for what is being carried is moving because of pulling or pushing or turning.

Carrying [is handled] thus. 'Turning, on the other hand,' he says, 'is composed of pulling and pushing.'[135] For the person who is turning [something] draws the thing being turned toward himself at one time, but pushes it away from himself at another, as in the case of those grinding grain with their hands, for the person produces the turning by pushing the millstone[136] away from himself and pulling it back again.

If, then, the four motions are reduced to pulling and pushing, but that which is pulling and that which is pushing are together with what is being pulled and pushed, it is evident that things imparting
1054,1 motion with respect to place are together with the things being moved by them with respect to place and nothing is in between them.

Accordingly, it remains to prove that what is pushing and what is pulling are together with what is being moved by them. He proves this, then, based on the definitions of pushing and pulling. The most authoritative type of demonstrations is that based on definitions.[137] The definition of pushing is motion either from a thing itself toward another or from one thing toward another, while [the definition] of pulling is motion from another thing either toward [the puller] itself or toward another thing, 'whenever', he says, 'the motion is faster than the [motion] separating the continuous things[138] from one another' [244a9-10]. The puller pulls the other thing along, either toward [the puller] itself or toward another thing, whenever, he says, the puller overpowers the proper motion of what is being pulled because the puller's motion is faster than that of what is being pulled. [The latter's motion], so far as depends on [the motion] itself, separates 'the continuous things from one another' [244a10], that is, the things which are pulled to [the point of] having touched and which are made continuous in this way. For the proper motion of each thing, so far as depends on [the motion] itself, separates and detaches the things which are joined [*sunaptomena*] by force with one another by the puller, for in many places Aristotle made a habit of calling things which are touching [*haptomena*] 'continuous'.[139]

But whenever the puller's motion is faster than the [motion] of what is being pulled which is separating the things about to be made continuous by the pulling [*holkê*], then pulling [*holkê*] occurs and the thing being moved 'is pulled along after' [244a11] by the mover. Or rather one should state [that it occurs] whenever the motion of that which is pulling a thing by force toward what is dissimilar [to it] and is detaching what up to now was continuous with what is similar [to it] is faster than the thing being pulled's natural motion separating the dissimilar continuous things from one another, for the natural powers making each thing continuous with what is similar [to it] keep things in their proper places, just as the [powers] which pull by force and detach things which up to now were continuous are pulling them to dissimilar places.[140] But even if they [pull] them toward similar [places], if, for instance, someone were to pull a stone downward or fire upward, still, the motion of what is pulling must be faster than that of what is being pulled, even if they occur in the same direction, if pulling [*holkê*] is going to be accomplished.

One ought to know that Alexander asserts that this text saying 'whenever the motion is faster than the [motion] separating the continuous things from one another' [244a9-10] is not transmitted in some manuscripts. But how could this make sense, since Aristotle immediately adds an objection deriving from this addition? However, in the other seventh book this text is as follows: 'There is pulling whenever the motion of the thing which is pulling, either toward it 1055,1 or toward a different thing, is faster, not being separate from that of what is being pulled.'[141]

244a11-b2[142] Perhaps there would seem to be a kind of pulling in another way too, {for wood does not pull fire in the manner described. But it makes no difference whether the puller is moving or stationary, for sometimes it pulls to where it is, but at other times to where it was. It is impossible for a thing to move something away from itself toward another or away from another toward itself without touching it, and consequently} it is evident that there is nothing in between what is imparting motion with respect to place and what is being moved.[143]

Having given [*apodous*] the account of pulling and in it having made [*apodous*] the motion of the puller faster than the natural motion of what is being pulled, he remarked that some pullers seem to be pulling though not themselves moving, as for instance a log seems to pull fire toward itself, and still more Heraclean stone [i.e. a magnet] [seems to pull] iron and amber [seems to pull] chaff, though not themselves moving. How, then, could the motion of these be faster than the natural motion of the things being pulled? [We may reply that] in the case of these things too that natural power pulling by force [i.e. in a direction contrary to the thing's natural motion] must be stronger than the power of the things being pulled, even if the pullers are not moving. This is why, he says, it makes no difference whether the pulling [*holkê*] occurs while the puller is moving or remaining stationary.

Explaining the [statement,] 'for sometimes it pulls where it is, but at other times where it was' [244a14], Alexander says, ' "Where it is" means "away from where it is", while "where it was" [means] "away from where it was" and [Aristotle][144] claims that a thing pulling away from itself toward another is drawing what is being pulled away from where the puller itself is, while [a thing pulling] from another toward itself is drawing what is being pulled away from where it was by drawing it toward itself.'

But the substitution of 'away from where' for 'where' seems unconvincing to me. Accordingly, [Alexander] explained [it] better, I think, further on: 'Having stated that it makes no difference whether <the puller is pulling while remaining stationary>[145] or moving, he added that a thing which is remaining stationary is pulling where it is itself, but a [puller] which is moving, having come to what is being pulled, draws that to its [i.e. the puller's] own place, in which it was before moving to what is being pulled.'

Alexander well inquires how it is by touching the things being pulled that things which pull naturally like Heraclean stone [i.e. a magnet] and such are pulling them,[146] and claims that even if they themselves do not touch them, nevertheless they do touch them with that power [or force, *dunamis*] by means of which they pull them. But the point proposed by Aristotle was not to prove that what is pulling touches what is being pulled with an incorporeal power, but rather with a corporeal one, so that there is no body in between.

1056,1 Accordingly, what is said next by him [sc. Alexander] is better: 'For either there are certain corporeal effluences from the things which are at rest and pulling in this way, by means of which, when they are touching and entangled, as some claim, the things being pulled are pulled, or else his present argument does not concern things which are pulled in this way (for what occurs in the case of those things is unclear), but he spoke about logs which, while remaining stationary, hold the fire down along themselves, somehow pulling it and forcibly overpowering its upward locomotion on account of being kindled from them.'[147]

Alexander remarked that in this passage Aristotle has added 'away from itself toward another' [244a15] to the text, which indeed is distinctive of pushing and pushing away.[148] But [we may reply that Aristotle] did not state both 'away from itself toward another or away from another toward itself' [244a15] about what is pulling, but rather 'away from itself toward another' about what is pushing, and 'away from another toward itself' about what is pulling. This is why, after having reduced the four motions to pushing and pulling, he next concluded generally that 'there is nothing in between what is being moved with respect to place and what is imparting motion'.[149]

Alexander asserts that after this text in some manuscripts some such text as the following is written: 'Similarly, even if there is something productive and generative of a qualified thing, it is necessary for this too to produce when touching, [as, for instance,] heavy [and] light.'[150] He asserts that immediately after this is found not the [statement], 'But, in fact, [there is] also not [anything in between] the thing being altered ...' [244b2-3], which indeed is written in most [manuscripts] after what has been stated earlier, but rather what comes a little after that: 'For in all [cases] it happens that the last thing causing alteration and the first thing being altered are together' [244b3-5].

'He might mean by such a text', [Alexander] says, 'that not only are the things imparting motion and being moved in the strict sense together, as for instance those causing change and being changed with respect to place and quality and quantity, but also those [causing change and being changed] with respect to form, that is, with respect to substance, which indeed[151] are coming to be and perishing: for in such things too what is causing change is together with what is being changed, as is the case when the heavy things and light things, that is, the corporeal elements, are coming to be.'[152]

Accordingly, the things producing and generating these are generative of some qualified thing, for the heavy and light are qualified things. For, in fact, in the next book he proves that these too are moved by something which has made them like that [sc. heavy or 1057,1
light].[153] But here he means that even if something is productive and generative of heavy and light, it is necessary for this too to produce [these] when touching, for [otherwise] the next [statement], 'it happens that the last thing causing alteration and the first thing being altered are together' [244b4-5], immediately seems otiose and shifted [from its proper location], since he talks next about what is being altered and causing alteration, and the text which will be discussed next is written in all the manuscripts.

244b2-245a11[154] But, in fact, [there is] also not [anything in between] what is being altered and what is causing alteration, {as is obvious on the basis of induction, for it happens that in all cases the last thing causing alteration and the first thing being altered are together. We start from the assumption that alteration occurs when things are affected with respect to the so-called affective qualities. It is owing to the number and degree of their sensible qualities that bodies differ from one another, and anything altered is altered by these qualities. This is so because these qualities are affections (i.e. variations in degree) of the underlying quality, for when a thing is heated and such we say that it is altered. We say this about animate and inanimate things alike, and in animate things about the parts which are not capable of sensation and the senses themselves, for the senses too are altered in some way, since actual sense perception is motion through the body when the sense organ is affected in some respect. Animate things undergo as many kinds of alteration as inanimate things do (though the reverse is not true since the latter are not altered with respect to the senses), but an animal need not always notice the affection. Since anything altered is altered by sensible qualities, in all these cases, at least, it is evident that the last thing causing alteration and the first thing being altered are together, for the air is continuous with the thing causing alteration and the body is continuous with the air. The same point applies to sight (the colour is continuous with the light and the light with the organ of sight); to hearing and smell (for the air is the first mover relative to what is moved); and to taste (for the flavour is together with the organ of taste). The claim that the last thing causing alteration and the first thing being altered are together likewise holds true for inanimate things and the insensate parts of animate things. Consequently there will be nothing in between} what is being altered and what is causing alteration.

After the things which are imparting motion and being moved with respect to place and, if indeed [he does discuss them], after the things [which are causing change and being changed] with respect to form,[155] he proposes to state, concerning what is being altered and what is causing alteration, that in the case of these too there is nothing in between, which indeed, he asserts, is known through induction, for in every alteration 'the last thing causing alteration and the first thing being altered' are found 'together' [244b4-5]. By 'last thing causing alteration' he means that which is proximately imparting motion and causing change, which up to now he was calling 'first'.[156] This is first in one way, as proximate to what is being moved, but last in another way, as the last of the movers, whenever the thing that is primarily [or first, *prôtôs*] and in the strict sense imparting motion is doing so by means of some intermediaries. By 'first thing being altered' he means that which is being altered in its own right and neither incidentally nor because any of the things belonging to it is being altered,[157] for that which is causing alteration is together with what is being altered in this way.

Having stated that it is 'obvious' [244b3] based on induction that

nothing is in between what is causing alteration and what is being altered, he next cites the [cases] through which one can grasp the point proposed on the basis of induction. Having proved first with an argument by what things the things which are altered are altered, he brings these [cases] before [our] view, showing inductively that the alteration occurs when these are together with the things altered by them. 'For', he says, 'it is hypothesized by us', that is, it is posited and agreed, 'that the things which are altered are altered by being affected with respect to the so-called affective qualities.'[158] We are acquainted with the [kind] involving the affective qualities as a species of quality in the *Categories*, this being fourth after state and capacity [*dunamis*] and shape.[159] Qualities are affective whose apprehension by the perceivers occurs through affection; of this sort are 1058,1
hotness, coldness, dryness, wetness, sweetness, bitterness, and such.[160] Also [included are those] conditions involving colours which are more superficial and not constitutive of form; all of which, he asserted, are called affective qualities not because they produce affection in the things apprehending [them], but rather because it is due to affection that they arise in the things acquiring them.[161] For the hotness of fire produces an affection in the things partaking [of fire], but, however, it is not spoken of as affective because it did not arise through affection so as to be a superficial and easily changeable condition; rather, it is substantial [*ousiôdês*].[162]

If, then, the things which are altered are affected, and the things which are affected are affected by the sensibles (for the things producing such affections are sensibles), therefore the things which are altered are altered inasmuch as they are affected by the sensibles.[163]

He asserts that sensibles are the things through which body differs from body.[164] [Bodies] differ in virtue of their qualities, for in fact even quantities are qualities in a sense. Accordingly, having body in common, [bodies] differ through [having] more or fewer qualities or by partaking of the same qualities to a greater and lesser degree.

If, then, the things causing alteration and being altered with respect to the affective qualities are bodies, it is necessary for them to cause alteration and be altered when adjacent to one another. For neither does a body do anything to another body separated [from it], nor is quality separable from body, nor is it active separately.

[Aristotle] stated, 'For these are affections of the *hupokeimenê* quality' [244b6], indicating that these are the affections belonging to the fourth species of quality hypothesized [*hupotethentos*], which is spoken of as 'affective'.[165] Or by '*hupokeimenê* quality' he means sensible [quality], which is what the argument concerns, because these are the affections of it.[166]

Next he divides the bodies which are altered into the animate and the inanimate, and the parts of the animate ones into those which are capable of sensation, such as flesh, sinews, and such; into the insensate, [such as] bones and hair and nails; and into 'the senses themselves' [244b10]. He seems to be taking the [parts] which are capable of sensation along with the [senses] when he speaks of 'the parts of animate [bodies] which are not capable of sensation and the senses themselves' [244b9-10]. 'For', he says, 'the senses too are altered in some way' themselves and not only the parts [*moria*] which are capable of sensation [244b10-11]. But they are not, in fact, altered in the way bodies are.[167]

He proved that the senses [*aisthêseis*] too are altered by stating how actual perceptions [*aisthêseis*] occur. For sense [*aisthêsis*] is threefold: the substantial [i.e. the sense organ], the capacity [*duna-*
1059,1 *mis*] of such a substance, and the activity [or actuality, *energeia*] of the substance in virtue of the capacity. Accordingly, this 'actual perception',[168] he says, such as seeing and hearing, 'is motion through the body' [244b11-12], when the part [*morion*] which is capable of sensation is being compressed [*sunkrinomenou*] or dilated [*diakrino-menou*].[169] But this is not yet perception (for insensate bodies too are compressed and dilated). Rather, whenever the sense as substance [i.e. the sense organ] itself suffers some appropriate affection, then actual perception is accomplished. He says what the affection is in the *de Sensu*, [namely,] that it is apprehension of the sensible form apart from the matter,[170] but an apprehension that is not purely active, as is intellect's,[171] which is from within and a whole,[172] but also involving something subject to being affected by the sensible. This is why it occurs along with alteration, and the substantial senses [i.e. the sense organs] as well are said to be altered.

Having stated how the senses are altered, [namely,] that it is because their activity involves something subject to being affected, next he gives the difference between the alteration of animate bodies and inanimate ones. First he speaks of the following [difference]: animate [bodies] are altered 'in just as many respects' [244b12] as inanimate bodies are, for they too are cooled and heated and moistened and dried, and both undergo all such alterations; but, in fact, an inanimate [body is] not [altered] in just as many respects as an animate one is altered, for nothing inanimate undergoes alteration 'with respect to the senses' [244b15]. He is not claiming here that all animate things are altered with respect to the senses, for of course plants, although themselves being animate too, are not altered with respect to the senses.[173] Rather, he is here speaking of as 'animate' those animals characterized by perception and by motion with respect to place.

He speaks of a second difference between the alteration of animate things and that of inanimate things: the inanimate ones never know nor perceive that they are being affected. On the other hand, animate things, that is, animals, sometimes do perceive that they are being affected, whenever they are affected with respect to the senses so that they see or hear or perceive with respect to some other sense, but at other times, even though they are capable of sensation, they do not perceive [that they are being affected]. For animals become paler and darker and increase and decrease [in size] though not perceiving these [affections].

Having stated these differences, he returns to the point proposed, bringing before [our] view [the fact that] the things altered by the sensibles are together. For, he says, 'In absolutely all ... it is evident that the last thing causing alteration and the first thing being altered are together.'[174] He proves that it is evident by citing instances of 1060,1
things causing alteration and being altered with respect to each sense.

First [he cites instances] with respect to touch. 'For', he says, 'the air' is 'continuous with the' sensible causing alteration, 'and the body' being altered '[is continuous] with the air' [245a5-6]. [He gives cases which involve an intervening medium] for if there were nothing in between [what is causing alteration and] what is being altered, the original [point] would be obvious. But also in cases where air is in between, he proves that this [air], which is the thing last, i.e. proximately, causing alteration, is together with the thing first being altered and is continuous with it, again calling that which is touching 'continuous' [245a5].[175]

Since the tangible is a sensible, he added touch too, which is not in some separated sense organ but rather in the body as a whole. For this reason he stated, 'and the body [is continuous] with the air' [245a6], for we perceive coldness and hotness and such things in this way, whenever what is causing the alteration is not immediately adjacent, but rather [operates] through air as a medium.

Similarly too in the case of the visible: 'The colour' which is visible is adjacent 'to the light' [245a6], that is, to the illumined air, for it is impossible to perceive anything visible without light. 'The light, on the other hand, [is adjacent] to sight' [245a7], that is, to the visual sense organ, for 'colour is capable of moving the actually transparent',[176] which indeed proximately moves the [organ of] sight.

In the same way both the organ of hearing and the organ of smell are adjacent to the first mover. He again speaks of the thing proximately imparting motion, which a short while ago he was calling 'last',[177] as a 'first mover' [245a8], having also added here the cause of its being spoken of as 'first'. For 'relative to what is being moved' [245a8] this is spoken of as 'first' in the sense of being proximate to it. It was demonstrated in the [books] *On the Soul* that hearing and smell occur through air.[178] In the case of taste, however, the tasteable flavour is together with the organ of taste, so there is no need for systematic discussion of the medium.

Having proved in the case of sensory alterations that the first mover and what is moved are together, he asserts that it will similarly be proved in the case of the alteration of inanimate [bodies] too that what is first, i.e. proximately, imparting motion is together with what is moved. For an inanimate body is neither heated at all nor cooled nor dried nor moistened nor in general affected in any respect, i.e. altered, if what is causing the alteration is not touching it.

If, both in the case of animate things and also in the case of inanimate ones, alteration occurs when what is proximately causing alteration is together with what is altered, it is safe to claim universally as well that nothing is in between what is causing alteration and what is altered.

After saying 'inanimate', he added 'and insensate' [245a10], either talking about the same things (for inanimate things are insensate too), or speaking of the parts [*moria*] in animate things which do not perceive as 'insensate'. Also, plants are insensate, but they are not inanimate.[179]

> **245a11-b2**[180] Nor indeed [is there anything in between] what is being increased and what is causing increase. {The first thing causing increase makes what is being increased larger by being added to it so that the whole becomes a single thing. Again, a thing which is diminishing gets smaller when something belonging to it departs. Accordingly, what is causing increase or diminution is necessarily continuous with what is being increased or diminished, and there is nothing in between things which are continuous. Hence, it is evident that} there is nothing in the middle of [what is being moved and the mover which is first, i.e. the one which is last relative to what is being moved].

After the things which are imparting motion and being moved in virtue of alteration, he proves that there is also nothing 'in the middle of' [245b2] what is proximately causing increase and being increased, nor yet [in the middle of] what is causing decrease and what is being decreased. These are more evident, since what is proximately causing increase, by being added to what is being increased and being united [with it], is increasing it in that way, and what is being decreased is diminishing and becoming smaller because something belonging to it flows away; what flows away was continuous with what is being decreased before flowing away.[181]

If, then, what is causing increase and what is causing decrease are what is being added and what flows away, and these were continuous in the strict sense with the things being increased and decreased, 'and there is nothing' in the middle of 'things which are continuous' [245a15-16], there is, therefore, nothing in the middle of what is causing increase and what is being increased and what is causing decrease and what is being decreased.

Having proved [it] individually in the case of each of the species of motion, next he concluded generally in the case of all 'that there is nothing in the middle of what is being moved and the mover [which is] first, i.e. [the one which is] last relative to what is being moved' [245a16-b2]. The thing proximately imparting motion is also spoken of as a 'first mover', viewed relative to what is being moved, because it is proximate to that. But it is spoken of as 'last' relative to it too, because the one which is proximate to what is being moved is the last of all the movers.[182]

[CHAPTER 3]

> **245b3-246a4**[183] [Based on the following one should consider] that everything altered is altered[184] {by the sensibles and that alteration can only occur in things which can be affected in their own right by the sensibles. The most likely other candidates are shapes and states and their acquisitions and losses, but none of those is alteration. When a thing has been shaped, we do not speak of it as that from which it has come but rather use a paronymous term derived from that; for instance, when a piece of bronze is shaped we do not speak of the resulting statue as 'bronze' but as 'brazen'. On the other hand, when a thing has been altered we do designate it as that from which it has come. Consequently, if what has come to be as the result of changes of shape is not spoken of with the same term but the product of alteration is,} it is evident that the comings to be [due to such reshapings] could not be alterations.

When proving that the motion involved in alteration occurs when the thing causing alteration and the thing being altered are together, he made use of [the argument] that alteration occurs in virtue of affection and that the things which are affected are affected by the sensibles, so that consequently the things which are altered are altered by the sensibles. Accordingly, having used that then, now he proves it. For he asserts that alteration is spoken of 'only in ... those things which' are affected 'in their own right ... by the sensibles' 1062,1
[245b4-5], meaning by 'in their own right' [*kath' hauta*] the things which are not being affected incidentally (for Socrates is visible

incidentally, but colour is in its own right). By [the phrase] 'by the sensibles' he means '[by] the affective qualities'. [This is] obvious, he asserts, from the [fact that] alteration occurs with respect to no other quality. For it is evident and agreed that alteration does occur with respect to the affective qualities. On the other hand, it not being evident that alteration occurs only with respect to these and with respect to no <other kind of>[185] quality, he proves this from the [fact that] neither changes with respect to shape nor those with respect to state are alterations.[186] Knowledge and virtue and health and sickness and such are states.[187]

Having stated that one might suppose there to be alteration 'in shapes and figures [*morphai*] and states',[188] he added: 'or[189] in the acquisitions and losses of these' [245b7-8], for having a quality is not the same as being altered, for alteration [lies] in change. Accordingly, he proves that there is alteration neither in shapes or states nor in their acquisitions and losses.

And first he produces the argument in the case of shape, proving that being shaped or changing with respect to shape is not being altered. 'Of the others one might above all suppose' [245b6], he says, that receiving a shape and a state is alteration, speaking either of all other things or, rather, of those in the [category of] quality. For there is another species of quality too, [namely,] capacity [*dunamis*], i.e. suitability, in virtue of which we speak of [people] as capable of boxing or capable of running.[190] He did not produce the argument concerning this species because it is not known whether such things are qualities and [the status of] the acquisitions and losses of them is still more unclear.[191]

He proves that a thing being shaped is not being altered by arguing syllogistically in the second figure as follows: A thing being shaped, when it has gained the shape and has been perfected [or completed, *teleiôthêi*] with respect to it, is no longer spoken of with the term with which the thing which was underlying and being shaped was spoken of, for the statue is no longer spoken of as bronze [*khalkos*] but paronymously as brazen [*khalkous*], and a bed is not spoken of as wood [*xulon*] but as wooden [*xulinê*], and a candle[192] [made] from wax is not [spoken of as] wax [*kêros*] but as waxen [*kêrinê*].[193] On the other hand, the thing which has been altered and has become hot or cold or has exchanged some such quality is still called by the same term;
1063,1 whether the thing which is heated or cooled or turns white or turns black or undergoes some such affection is bronze or wood, it remains being spoken of as bronze and wood after the affection too, for it is spoken of as liquid or hard bronze and wood.[194] And not only this, but also conversely 'we speak of' this 'liquid as bronze', if it happens to have become molten, and 'the hot thing' as wood [245b15-16]. Conse-

quently, from wherever we begin,[195] a thing which has been altered is said to be the very same thing which it was earlier too.

He added conversion[196] so that, even if one were forced, in the case of things which have been shaped, to call them by the same term when beginning from the underlying thing, [saying that,] for instance, the wax is a triangle or the bronze is a circle, one would no longer be able to keep the same term when beginning from the shape, for it is not possible to say that the triangle is a wax or the statue a bronze.[197] But [in the case of alteration we] designate the matter 'homonymously with the affection',[198] he says, in that we speak of both the affection and also the bronze as 'liquid' and the affection and the wood as 'hot'.

Having established the premises by means of several [cases], then having set them out succinctly, and first the minor [premise], when he says, 'Consequently, if with respect to shape' and so on [246a1-2], then the major, when he says, 'But, with respect to affections and alterations, it is said [to be that]' [246a2-3], thus he draws the conclusion.[199]

The point proposed was to state that shapings are not alterations, but he inferred that comings to be are not alterations [246a3-4], showing whence the argument has [its] strength, [namely,] that it is based on the [consideration that] reshapings are already comings to be of some kind. It has been proved in the fifth book that comings to be are not motions,[200] and consequently not alterations either, for alterations are motions of some kind. Accordingly, he reminds us of the things proved there, for if a thing which is altered is altered while remaining just what it is, as it indicates by keeping the same term, whereas a thing which is coming to be is changing and not remaining just what it was, [it follows that a thing which is coming to be is not being altered].[201] But a thing being shaped also does not remain completely just what it was, as not being called by the same term indicates, even if it is not coming to be completely. For a thing being shaped seems [*eoike*] to have a nature intermediate between things which are coming to be and things which are being altered, but it seems [*dokei*] to him to incline more toward a thing which is coming to be, because it is not called by the same term either, which indeed pertains to things which are coming to be. However, being referred to paronymously indicates its diverging from things which come to be completely.

Someone might pose a puzzle perhaps, [fearing] that affective
qualities, by penetrating deeply, might lay hold of the underlying 1064,1
thing more, and change it more, than a shape which has disposed the surface alone. [We may reply that] even if affective qualities do somehow penetrate deeply, nevertheless they are adventitious and are incidental features [or accidents, *sumbebêkota*], producing some

faint and easily lost tincture,[202] for it is with respect to such [qualities] that alteration occurs. For instance, iron is altered when heated superficially, but if it has been heated red-hot, this is no longer alteration, but passage to a different form [or species, *eidos*]. Hence it then has fiery activities [i.e. behaviour and effects] too.[203]

> **246a4-9**[204] Further too, [it would seem] absurd to state that {a human being or a house or any other of the things which have come to be has been altered, though perhaps it is necessary that each of them comes to be when something is altered, for instance when the matter is condensed or rarefied or heated or cooled. However, the things which come to be are not altered,} nor yet is their coming to be alteration.

Having proved that in the case of changes with respect to shape the underlying thing and the thing which has been shaped are not similarly spoken of, and having in this way assimilated things which are shaped to things which come to be, he proves in the case of coming to be that coming to be is not alteration, since he has that as the remaining [point in his proof] that reshaping too is not alteration. For it is 'absurd' [246a4], he says, to claim that a human being who has come to be or a house which has come to be has been altered, for they have come to be some other [*alla*] things and not [merely] otherwise [*alloia*],[205] for the change has occurred with respect to substance [or essence, *ousia*], not with respect to something incidental to them. For a thing being altered and remaining just what it was even after having been altered is being altered in this way,[206] but a thing which is coming to be is coming to be when it does not yet exist, for neither the human being nor the house existed yet. On the other hand, that which was [in the process of] being altered is also what has been altered.

Having stated that it is 'absurd' to claim that that which has come to be has been altered, he rightly (since alteration does appear in coming to be and also nothing could come to be if alteration did not exist) makes the following distinction, saying that 'each' of the things which comes to be 'comes to be' [246a6] when the underlying matter is altered. For that endures the changes while remaining just what it is, but the coming to be of that which comes to be is not in fact alteration, for what is not yet existing cannot be altered, since a thing which is coming to be does not yet exist.

246a10-b3[207] But, in fact, nor yet are states, neither those of the body [nor those of the soul, alterations].[208] {Some states are virtues, others vices, but neither virtue nor vice is an alteration. Rather, virtue is a kind of perfecting (*teleiôsis*) (for it is when it has acquired its virtue that a thing is spoken of as perfect, since then it is most in accordance with nature), whereas vice is a perishing of and departure from perfection. Accordingly, just as we would not speak of the completion (*teleiôma*) of a house as an alteration, we also should not do so in the case of virtues and vices and those having or acquiring them. The virtues are perfectings, the vices departures from perfection,} and consequently [they are] not alterations.

Having proved that change with respect to shape and figure is not alteration, but rather coming to be, he proves that change with respect to state, which is the first species of quality,[209] is also not alteration, but rather coming to be. He divides states into the psychic 1065,1
ones, like virtue and vice, and the bodily ones, like health and sickness, none of which is sensible, in order, by proving that these too are not alterations, to strengthen the [claim that] alterations occur with respect to the affective, i.e. sensible, qualities alone, which indeed was the point proposed.

'Nor yet are states', he says, that is, changes with respect to states, 'alterations' [246a10-11]. [This is] because states do not come to be present through alteration. He proves that they are not alterations by reducing all [of them] to virtue and vice, for in fact health and sickness too are virtue and vice of the body. If, then, a thing changing to its own virtue is changing to its own natural condition;[210] and a thing changing to its own natural condition is then gaining [its] own proper perfection; and when each thing has gained [its] own proper perfection, then it is just what it is; and when it has proceeded to being just what it is, then it has come to be but has not been altered – and the conclusion is obvious.[211]

He himself [gave the] conclusion, after stating: 'Whenever it has acquired its own virtue', 'then' it is 'most in accordance with nature'. When it is 'most in accordance with nature', 'then each thing is spoken of as perfect'.[212] A thing gaining its own perfection cannot be said to be altered, for the house which is being roofed with tiles is also not said to be being altered, but rather to be coming to be. Just as the change to each thing's own proper virtue is a kind of coming to be of it [sc. of the thing], so too the [change] to [its] own proper vice is 'perishing' [246a16].

He admirably cited the example of the circle, indicating that just as some things are spoken of as circles, though, however, they are not circles if they do not have perfection [*to teleion*], so too in the case of absolutely all other forms anything not having its own perfection [*teleiotês*] is not in the strict sense that which it is spoken of as, for a thing which is imperfect [or incomplete, *ateles*] with respect to form does not admit the definition of the form either.

If changes to the virtues are perfectings, whereas [changes] to the vices are perishings of and departures from perfection, changes to states would, rather, be comings to be and perishings, not alterations. For it would be absurd to speak of laying round the coping or putting on the roof tile as alteration of the house which does not yet exist but
1066,1 is still coming to be. Similarly too, change to virtue is perfecting, and the perfecting contributes to coming to be.

But one might inquire, I think, how it is possible to speak of virtue and vice as the coming to be and perishing of the soul, which indeed, having the form of the human soul and abiding in the same [form], becomes virtuous at one time but corrupted at another. This is why the same person becomes wicked at one time but excellent at another. How is the acquisition of virtue similar to that of roof tile or a circuit wall? For, whereas the latter are parts of the house and not states, the former are not parts, since if virtue were a part of the soul, the [soul] which has lost [its] virtue would have perished [*apolôlei*].

[The solution is that] in general, perfection is of two kinds: on the one hand that of the substance itself, in virtue of which it is completed by the first, middle and last parts of itself,[213] though [this perfection] is not itself a state (for what would the possessor be, inasmuch as the form does not yet exist apart from such perfection?); on the other hand, the [perfection] involving virtue and vice and, speaking generally, involving the state is extrinsic[214] to the whole form and supervening, for in fact these come to be and go out of being apart from the perishing of the underlying thing.

How, then, is it that as an example of a state he cited the perfection involving the parts as in the case of a house? Perhaps, then, each form which is in accordance with nature is completed not only by the perfection of [its] own proper parts, but also by [its] own proper virtue, for to be in accordance with nature is nothing other than to have [its] own proper virtue, so as to perform the activities in accordance with nature. And for this reason Aristotle syllogistically deduced perfection from [being in] accordance with nature. Just as, then, a body which is sick could not be perfect, because it is not able to display the activities in accordance with nature, even if it does have all the bodily parts (for it has them as dead ones, given that they have been deprived of [being in] accordance with nature, and it is like a corpse), so too a rational soul, when it has lost the virtue naturally befitting it and is not able to perform the activities naturally befitting it, is not a soul in the strict sense, nor yet a rational life[215] in accordance with nature, but rather a deadened kind of one.

One must, then, reckon in addition as the most important part of the whole substance of each thing the virtue naturally befitting it, apart from which the rest is dead and spoken of homonymously.[216] Not displaying the proper activities belonging to the form [or species, *eidos*] indicates [this]. Aristotle, then, rightly compared the perfection involving the state to the parts of the house on the grounds that it is the most important part. For the state, i.e. the virtue, is also not, in truth, an incidental feature (since it destroys the underlying thing 1067,1
when it withdraws), just as the soul is also not an incidental feature of the animate body.

> **246b3-20**[217] Further, we also assert that absolutely all the virtues [lie in being somehow relative to something]. {The bodily virtues such as health and good condition depend on the mixture and proportion of things either within the body or relative to the environment, and similarly with the other bodily virtues and vices. Each is a relative and disposes its possessor well or badly toward the affections through which the thing comes to be or perishes. Accordingly, since neither are relatives themselves alterations, nor is there alteration or coming to be or any change at all of them, it is evident that neither states nor their losses and acquisition are alterations, though perhaps it is necessary for them to come to be and perish when certain things are altered. Each vice and virtue is spoken of in connection with the things by which its possessor is naturally altered. Virtue makes a thing either unaffected or subject to being affected in a certain way, but vice makes it either subject to being affected} or unaffected in the contrary way.

He proves that changes with respect to the virtues and the vices are not alterations on the basis that they are not qualities but are classified under the relatives; he himself has proved in the fifth book that neither alteration nor any change at all occurs with respect to the relatives.[218] He proves that the virtues and the vices [fall] under the relatives by making the argument first in the case of [the virtues and vices] of the body, then passing from those to the ones of the soul. For, he asserts, health and good condition, which are virtues of the body, [lie] in due proportion of the bodily qualities and the primary powers, [namely,] hotness, coldness, dryness, wetness,[219] when they have due proportionality either relative to one another or also 'relative to [their] surroundings' [246b6], so that on account of such due proportion the bodies are not easily mutable and easily affected by [their] surroundings. Beauty too, which also[220] is a virtue of the body, [lies] in the due proportion above all of the nonhomoeomerous parts and the complexion. And strength, which itself[221] is also a virtue of the body, [lies] in the due proportion of the homoeomerous [parts], above all of sinews and bones.[222] But due proportion is a relative. Accordingly, 'each' of the bodily virtues [lies] in 'being somehow relative to something' [246b8], since it is due proportion. Similarly too, the vices opposed to the virtues [lie] in disproportion of the same affections, for [they lie] in excess or deficiency.

Also, each virtue 'well disposes' the [person] having it 'in connection with the proper affections' [246b9]. The 'proper affections' are the ones 'by which' [246b10], when duly proportioned, [bodies] come into being and are in good condition, [namely,] hotness, coldness, dryness, wetness,[223] and such, but, when disproportioned, bodies perish and become corrupted, becoming more easily affected and easily changed in their proper affections by [their] surroundings and by one another.

If, then, the virtues and vices of the body [lie] in being somehow relative to something, and it has been proved in the fifth book that in
1068,1 the [category of] relative there is no motion in its own right, nor any change at all,[224] it is obvious that there could not be alteration either. For every change was proved[225] to occur either with respect to substance, like coming to be and perishing, or with respect to quantity [*poson*], like increase and decrease, or with respect to quality [*poion*], like alteration, or with respect to place, like locomotion. If, then, the virtues and the vices of bodies are in the [category of] relative as instances of due proportion and disproportion, but relatives are not in [the category of] quality [*poiotês*], and change of things which are not in [the category of] quality [*poion*] is not due to alteration, it is obvious that change with respect to the virtues and the vices is not alteration.

Having proved that the virtues and the vices and in general the bodily states are in the [category of] relative, he adds that 'relatives are neither themselves alterations' [246b11], like heating, so that the thing partaking of them is altered. For, whereas we do say that the thing which is heated or cooled is altered, we do not say that the thing which comes to be double or on the right or duly proportioned or similar or equal is altered. Furthermore, there is no alteration of relatives, just as [there is none] of substance, so that one would not ever say that relatives which come to be and acquire a relationship to something are altered too, for a double magnitude which comes to be equal is not altered in the way a thing being heated is. The essence of alteration[226] differs from the thing being altered, in that alteration is partaken of by the thing being altered, as heating is by the thing being heated, whereas the thing being altered is altered through partaking of alteration. Accordingly, a relative, he says, is neither alteration nor is it altered. He proved in the fifth book that there is neither coming to be nor motion nor any change at all with respect to relatives on the basis that something which does not change in any respect itself is double at one time and not double at another because another[227] thing changes.[228]

'It is evident' [246b12], then, from what has been stated earlier that change with respect to shape is not alteration, and from what [has been stated] now that [change] with respect to states is not either. Therefore, only change with respect to the so-called affective qualities is alteration.

Having proved that changes with respect to the virtues and the vices of the body and in general 'the losses and acquisitions of states' [246b13] are not alterations, next he teaches the cause why things changing with respect to them seem[229] to be altered, and he claims that 'it is necessary' for them to come to be and perish 'when certain things are altered' [246b15], for health comes to be when certain things belonging to the body are heated and cooled, that is, when they are altered. But neither is the alteration the health, nor the health the alteration; rather, due proportion of the things being altered, owing to which the health is observed, supervenes upon the alteration. The same account [applies] both in the case of strength and in the case of beauty, for if these supervene when certain things are altered, how could anyone claim that things which supervene and are not yet existing are altered? For something changing to health is not being altered just because of that; rather, it is changing through being altered. If being healthy [lay] in being hot or cold, the change to health would be alteration;[230] but since health is none of those and has [its] essence [*einai*] in the due proportion of them, it is obvious that the change to health is not alteration, though it supervenes through alteration.

He said 'just like the form [*eidos*] and the figure [*morphê*]' [246b15-16] to indicate that just as coming to be (for this is what change to 'the form and the figure' is) is not alteration in virtue of the acquisition and loss of the form, though it occurs when certain things are altered, so also in the case of change with respect to state, for this too is not alteration, though it supervenes when certain things are altered. Or, [alternatively,] he said 'the form and the figure' to indicate shape and change with respect to that, for he said in the case of that as well that 'perhaps each comes to be ... when something is altered, for instance when the matter is condensed or rarefied' [246a6-8].

Having stated that states arise when certain things are altered, 'for instance, hot and cold ... dry and wet' [246b16-17], he added 'or the primary things in which they [i.e. the states] happen to exist' [246b17], since it is not proposed by him to inquire now about what things the affections belong to and in virtue of what primary changes and alterations states arise, [that is,] whether [they belong to] the things which are homoeomerous, as Anaxagoras [claims],[231] or the atoms, as Democritus does, or the qualities, as Aristotle does.[232] For it is quite obvious that [states] arise when certain things are altered, but let just what things [these are] be inquired about [on another occasion].

He confirms the argument also from [the consideration that] vice and virtue are spoken of in connection with the affections 'by which the possessor is naturally constituted to be altered' [246b18-19]. 'For virtue' when it arises 'either makes' [its possessor] completely 'unaffected' (if, at any rate, we call only the bad ones affections)[233] 'or subject to being affected in a certain way' [246b19-20], that is, moderately, if we consider virtue [to lie] not in impassivity, but rather in due proportion of the affections. For in the case of the soul the moral
1070,1 and the civic virtues are of this kind, not eradicating the emotions[234] and the appetites completely from the soul, but moderating their movements. In the case of the body as well, bodily virtue, that is, health, does not make the body unaffected by hot and cold, but rather moderates the affections arising from them, just as [bodily] vice, i.e. sickness, is lack of due measure of them.

Accordingly, if the virtues and the vices involve these things by which and with respect to which the thing possessing them is naturally constituted to be altered, and alterations are with respect to the affective qualities, [then] the virtues and the vices of the body would involve the affections and the affective qualities of the body, so as to arise when those things are altered. Since it is not [the subject] proposed, he here omitted inquiring about which affections each of the virtues of the body, such as health and strength and beauty, involves. Just as virtue makes [people] unaffected by the bad affections but subject to the moderate ones (which indeed some do not deem it proper to call affections),[235] in the same way vice makes [them] subject to the bad affections but unaffected by the moderate ones. Consequently, virtue and vice involve affections and arise through alteration with respect to them, virtue through due proportion [or due measure, *summetria*] of them, vice through lack of due measure [*ametria*]. Consequently, neither states nor changes with respect to states are alterations.

Since in this [passage] too Aristotle claims that there is no alteration nor coming to be nor 'any change at all' [246b12] of relatives, one must remark that, though having stated that the virtues and the vices are among the relatives ('for each', he says, 'exists through being somehow relative to something' [246b8]), he claims that they come to be and perish when certain things are altered. How is it, then, that there is no coming to be nor any change at all of relatives? [We may reply that] relatives too assuredly do come to be and perish, for a thing

which is not double earlier is [double] later and then again is not, and anything which, after not being earlier, is later comes to be, and anything which, after being earlier, is not later perishes. But they do not come to be and perish in their own right; rather, as he specified here, these things come to be and perish when certain things are altered or, speaking generally, when they change. 'For it is possible', he says in the fifth book, 'when one thing changes, for another which is not changing at all to be truly described by a predicate, so that their motion is incidental.'[236] In that [passage] he dismissed incidental 1071,1
change as indefinite. 'For', he says, 'it [occurs] in absolutely all [respects] and always and belongs to everything.'[237] Accordingly, if relatives do not exist in their own right but have [their] being in other things, then, when those things change, the relatives change too, not in their own right but incidentally.

> **246b20-247a19**[238] Similarly too in the case of the states of the soul, {for all of them too are somehow relative to something, and the virtues are perfectings whereas the vices are departures from perfection. Further, virtue disposes a person well, and vice badly, relative to the proper affections. Consequently these too are not alterations. Nor are the losses and acquisitions of them alterations, though they necessarily come to be and perish when the sensory part is altered. This alteration is brought about by the sensibles, for all moral virtue involves bodily pleasures and pains, and those occur in the course of doing something or remembering or expecting. The pleasures and pains involved in action are due to sense perception and hence are caused by some sensible, and those involved in remembering or expecting are derived from this, for people are pleased when they remember what they have experienced or are looking forward to what they are going to experience. Consequently it is necessary that all such pleasure is brought about by the sensibles. Since it is when pleasure and pain arise that virtue and vice arise, and since all such pleasures and pains are alterations of the sensory part, it is evident that it is necessarily when something is altered that they are lost and acquired. Consequently, their coming to be is accompanied by alteration,} but it itself is not alteration.[239]

Having proved that the virtues and the vices of the body and in general the bodily states do not come to be and go out of being through alteration, since virtues are perfections in accordance with nature whereas vices are contrary to nature, and things proceeding to [being in] accordance with nature are coming to be and not being altered, and those [proceeding] to being contrary to nature are perishing (even if they are in the [category of] relative), he passed next to the states of the soul. He proves that changes with respect to these are not alterations either, since these too, just like the bodily [states], are perfections of the soul and have [their] existence [*hupostasis*] in being 'somehow relative to something' [247a2]. Accordingly, acquisitions and losses of them are rather, as it were, comings to be and perishings of perfection, not alterations, for each of the things that exist, whenever it is perfect and in accordance with nature, is then most of all the very thing it is said to be, for in fact a [horse] possessing the virtue of a horse is a horse most of all, and each of the other things too.[240]

It is obvious that the virtues of the soul are also relatives, since they too are instances of due proportion of those things in which they arise – emotions [*thumoi*] and appetites and such – just as the bodily [virtues are instances of due proportion] of hot and cold [bodily constituents] and such. He himself indicated their relativity by the [statement]: 'Virtue disposes [its possessor] well relative to the proper affections, vice badly' [247a3-4]. For just as it pertained to the bodily states, which are relatives, that each of them disposes 'the [body] possessing' it 'well or badly in connection with the proper affections' [246b9], and [just as bodily] virtue, through due proportion, and vice, through disproportion, make [bodies] 'either unaffected or subject to being affected in a certain way',[241] so too the [psychic] virtues well dispose the possessors 'relative to their proper affections'[242] and the vices 'badly' [247a3-4], so that these too are relatives, being instances of due proportion and disproportion.

Accordingly, if neither are relatives themselves alterations nor is
1072,1 there alteration of them, then neither are psychic states nor the losses and acquisitions of them alterations. However, 'it is necessary' for these too [i.e. psychic states], just like the bodily ones, 'to come to be when' something 'is altered' [247a6]. For bodily states come to be 'when certain things are altered' [246b15] 'in which they happen to be primarily',[243] [namely,] hot and cold [bodily constituents] and such, whereas psychic [states come to be] 'when the sensory part is altered' [247a6-7]. This 'will be altered', obviously, 'by the sensibles' [247a7] to which it is naturally related. He said 'when the sensory [part] is altered' on the grounds that it is what is subject to being affected, for the affections, as Plotinus says, are either perceptions or not without perceptions.[244]

Having first stated generally concerning psychic states that they too are 'perfectings' and in the [category of] relative,[245] since they are certain due proportions and disproportions, and that they come to be 'when the sensory [part] is altered' [247a6], he next divides the psychic virtues into the moral ones and the intellectual ones and proves in the case of each how they are in the [category of] relative and how neither is an alteration and there is also no alteration of them, but they come to be when certain things are altered.

He proves that moral virtue comes to be when the sensory [part] is altered on the basis that every moral virtue 'involves bodily pleasures and pains' [247a8], as is demonstrated in the *Ethics*.[246] Bodily pleasures and pains come to be when the sensory [part] is altered by the sensibles, and consequently moral virtue comes to be when the sensory [part] is altered by the sensibles.

He proves by division that pleasures and pains come to be when the sensory [part] is altered. For we are pleased or pained 'either in doing' something 'or in remembering' (for in fact the memory of pleasant things produces pleasure and that of painful things [produces] pain) 'or in expecting' (for in fact expectations of pleasant things also please, while those of painful things cause pain) [247a8-9]. If, then, the pleasures and pains 'in doing' are sensible and are produced [*ginontai*] by sensibles, and those produced 'in remembering' or 'expecting' are also produced by sensibles ('for' we are pleased or pained 'either when we remember the kinds of things which we underwent' [247a12] and which we perceived or when we are thinking of the kinds of things which we will undergo and the kinds of things we will perceive), it is obvious that pleasures and pains result for us when the sensory [part] is altered by the sensibles.

Though it is proved, as I stated,[247] in the *Ethics* through several [arguments] that moral virtue involves pleasures and pains, let it be 1073,1 said concisely here as well that habituation in being pleased and pained – not at random, but in the things one ought and when one ought and as much as one ought[248] – produces moral virtue. Plato too teaches this concerning pleasure and pain in the *Laws* when he says: 'For these two springs have been set free to flow, from which the person who draws whence and when and as much as one ought is happy, and so with absolutely every city and animal.'[249]

Having demonstrated the premises, [Aristotle] sets out the syllogism succinctly thus: Virtue and vice of soul arise 'when pleasure and pain arise' [247a14-15], and pleasures and pains are 'alterations of the sensory [part]' [247a16-17]; 'it is evident', therefore, that virtue and vice arise and perish when the sensory [part] is altered, so as 'to lose' them 'and to acquire' [them] at that time [247a18]. It is obvious, then, from what has been stated that 'the coming to be' [247a18] of virtue and vice follows when certain things are altered, 'but' their coming to be 'itself is not alteration' [247a19]; rather, it supervenes upon the alteration of the sensory [part] involving pleasure and pain, virtue owing to due measure arising through this alteration, and vice owing to lack of due measure.

He said, 'Pleasures and pains are alterations of the sensory [part]',[250] instead of 'come to be when the sensory [part] is altered'. [I draw attention to this point] because the alterations[251] of the sensory [part], for instance, heating or cooling or dilating or compressing[252] or however one might be able to name their individual types, are affections of the sensory [part] arising under [the influence of] the sensibles. Upon these [affections] there supervenes sometimes pleasure, sometimes pain. When [the pleasures and pains] which arise are duly proportioned, virtue follows, but when they are disproportionate, vice does.

> **247b1-7**[253] But, in fact, nor yet [is] the state[254] of the intellectual part {alteration, nor is there coming to be of it, for we say that a knower above all is in the category of relative. Further, it is evident that there is no coming to be of such states, for a potential knower becomes an actual knower not because it has changed but because some other thing exists: whenever the particular has come to be, the intellect} somehow knows the particulars through the universal [knowledge].[255]

Having proved that the moral virtues and vices and their acquisitions and losses are not alterations, he proves next that neither[256] the
1074,1 [virtues] belonging to the intellectual part nor their acquisitions and losses are alterations. Just as the virtues connected with the irrational [part] of the soul, in which there are pleasures and pains, are 'moral' (these belong to irrational animals in common [with human beings], for among them too many things are rightly done through habituation),[257] so [too] the [virtues] connected with the rational [part] of the soul, where knowledge comes to exist and the virtue with respect to it, are 'intellectual' [*noêtikai*] or 'rational' [*dianoêtikai*] (for it is written either way).[258]

Accordingly, he says, 'Nor yet' is 'the state' of this [part of the soul] 'alteration', nor is there alteration or 'coming to be of [the state]' [247b1-2], since he reduces this too to the [category of] relative, for he stated a short while ago concerning relatives that 'neither are they themselves alterations, nor is there alteration of them, nor coming to be, nor any change at all',[259] obviously [meaning] in their own right, since he wishes relatives too to come to be and change incidentally. If, then, the knower is in the [category of] relative and there is no alteration of a relative, there is also no alteration of the knowing state,[260] nor coming to be, nor any change at all in its own right.

Establishing [this] on the basis of an *a fortiori* [argument], he mentioned to begin with that the state involving knowledge is in the [category of] relative, pointing out that the knower is said to be 'in [the category of] that which is somehow relative to something' [247b2-3][261] much more than moral states are. For knowledge possesses relativity more evidently than do the virtues in the moral [part, such as] temperance and justice. Whereas those are reduced to the [category of] relative because of the due proportion [involved in them], knowledge is of a knowable and all cognition is of the cognizable.

Second, he proves that the states of the intellectual [part] are in the [category of] relative through the [fact that] there is no coming to be of them. For it is distinctive of relatives that the potential in them comes to be in actuality though the thing itself does not change at all in its own right. For one must always remember that Aristotle denies coming to be and change in their own right to relatives, though admitting that they[262] do come to be and change incidentally. However, it is obvious that anything which does not change at all itself in its own right does not come to be in its own right either, for every coming to be is a change.

Accordingly, if 'the potential knower' [247b4] in the intellectual [part] comes to be and comes to exist in actuality, then at that time the knowledge, which is cognition of the universal, [comes to be and exists,] whenever on the basis of perception an experience [consisting of] cognitions of the individuals has been brought together.[263] For we know scientifically that every human being is rational from inspection [*ephodos*] and trial [*peira*] of the individuals, which does not occur 1075,1
in the intellectual [part], but in the sensory and imaginative ones. Accordingly, when the trials of the particular have been brought together in them,[264] the potential knower in the intellectual [part] has come to be in actuality, 'though not having been moved at all itself' [247b4-5].

This, then, is the way Alexander explained [it], wishing the universal and the cognition of the universal to be brought together from the particulars,[265] and he said that it was stated that 'it somehow knows the particulars through the universal [knowledge]'[266] as a sign that the cognition of the universal is gathered by means of the particulars, since the universal knowledge is of each of the things [falling] under the universal (and it has been gathered from these), for the particulars are comprehended in the knowledge[267] of the universal, on the grounds that that [knowledge] comes to be from them and in dependence on them.

On the other hand, if it is not possible for universals to be brought together from particulars, since those are infinite, nor for the cognition in the sensory and imaginative [parts] to be capable of bringing knowledge into existence in the intellect, [this cognition] being much inferior to [knowledge, which is] superior, one must rather, I think, explain in a simpler and truer way what has been stated by Aristotle. The intellect always has knowledge of the universals in actuality (whether ready to hand or not),[268] but of the particulars [only] potentially. Whenever perception strikes the particular, at that time, then, the intellect actually knows [*ginôskei*] the particulars through the universal. 'For an individual human being is a human being too', as [Aristotle] himself stated elsewhere.[269] For this reason too he added the 'somehow' [*pôs*, 247b6], because the cognition of the universal is not on the same level as that of the particular; rather, the particular is known [*ginôsketai*] 'through the universal [knowledge]'[270] because the particular is encompassed by the universal.

247b7-9[271] Again, [there is no coming to be] of using and activity {unless it is thought that there is some beginning of seeing and touching,} for using and being active are similar to these.

He has proved that the potential knower of what is particular and sensible, though not moved at all itself, when the particular has come to be and has come to perception, knows [*ginôskei*] it through the universal [knowledge], from which it has been inferred that [the knower] is a relative and for this reason there is neither alteration nor coming to be nor any change at all of it in its own right. Now he
1076,1 proves that if the knowable thing exists, whether particular or universal, then whenever the knower is active in connection with it, using the knowledge, such activity is not accomplished through coming to be, nor is there coming to be of such using. There is no alteration of it for the [same] reason there is no coming to be of it: there is no coming to be of it because it is an activity, for coming to be is in time, but the change from not being active to being active occurs atemporally, just like both seeing and touching.[272] That the change to these occurs atemporally is obvious because a time is found which is less than every time hypothesized as the least in which the change from not touching and not seeing to touching and seeing occurs. Consequently, such a change could not be alteration either, for alteration too is in time.

Further, if seeing and touching are activities, whereas alteration is an affection, there could not be an affection of the activity insofar as it is an activity. Similarly, the change from not being active with respect to knowledge to being active also does not occur through coming to be, and not through alteration either.

247b9-13[273] The original acquisition {of knowledge is not coming to be, and not alteration either, for it is because the mind is at rest and has come to a halt that we are said to know and to understand, and there is no coming to be of coming to rest,} as has been stated earlier.

Not only do the using and the activity in virtue of the knowledge which is present as a state not change from not being active to being active through coming to be or through alteration, but [the same is true of] 'the original acquisition of the knowledge' [247b9-10]. This is the change from the first potentiality to the state which is able to be active in the future, in virtue of which, though we cannot yet have knowledge, we change to being able to have knowledge in the future.[274] A settling down of this kind occurs after the first stage of life when the mind comes to rest and quiets down from the great disturbance which it endures on account of the many accretions and excretions of the body occurring due to nutrition and growth, when the nature is still rather weak. He asserts, then, that such a change is neither alteration nor coming to be. For coming to rest, that is, change to rest, is motion, as has been proved earlier,[275] and 'there is no motion of motion, nor coming to be of coming to be, nor any change of a change at all',[276] as has been proved before and is brought to mind here with the [statement]: 'For there is no change of coming to be, as 1077,1
has been stated earlier.'[277]

Aristotle nicely derives 'to know' [*epistasthai*] from 'the mind has come to a halt' [*stênai tên dianoian*],[278] here too voicing [views] in harmony with his master. For in the *Timaeus* Plato also asserts that these are the causes of the original lack of intelligence of human souls and that the gaining of knowledge is like this. For, after the creation of the things which are everlasting, the [gods] who were setting mortal things in order (he says) 'bound the revolutions of the immortal soul in a body which is subject to inflow and outflow. Bound in a great river, the [revolutions] were neither dominating nor dominated, but were carried along and carrying along by force'.[279] After stating many things concerning such settling down, he also adds: 'Accordingly, on account of all these affections now and at the beginning, at first when it has been bound in a mortal body the soul comes to be lacking intelligence, but when the stream of growth and nutrition comes on less and the revolutions, acquiring calm again, go their own way and settle down more as time goes on, just then the rotations of each of the circles straighten out as they travel toward the shape which is in accordance with nature.'[280] For if even in adults the motions of the body and those of the irrational desires become causes of folly, how much more so in the young, in whom the bodily movements are even more vehement, the nature weaker, and reason both untrained and inexperienced? Plato adds the following too: 'Accordingly, if a correct nurture also takes part in the education process, [a person,] perfect [*holoklêros*] and totally healthy, escapes the greatest sickness; but one who is neglectful[281] of the way to live, having passed through a lame life, comes again to [the house] of Hades imperfect and unintelligent.'[282]

The whole argument is formulated by Aristotle as follows: 'The original acquisition of knowledge' [247b9-10] occurs through coming to rest; there is no motion of coming to rest, and no change either, since coming to rest is a change and there is no change of a change, for there would be something changing to opposites simultaneously and arriving at opposites simultaneously. And further, the procession
1078,1 of change and coming to be will be to infinity, as has been proved earlier.[283] For these reasons, there also could not be a change of coming to rest.

He said 'on the one hand [*men*] is not coming to be'[284] as though going to give something by way of contrast, but he plainly does not give anything by way of contrast, unless perhaps what is stated after many [other] things, [namely,] 'on the other hand [*de*] in both cases [this occurs] when certain of the things in the body are altered' [248a3-4], is given in answer to this too, as Alexander asserts. Perhaps when he found the [following] text, which has 'for it is because the mind is at rest and has come to a halt ...' [247b11], Alexander believed the sentence [*logos*] to be defective, on the grounds that there is no 'on the other hand' given in answer to the 'on the one hand'. But, first, the 'for' [*gar*] stating the cause is given in some way, I think, in answer to the 'on the one hand'. Second, in some of the manuscripts it is written 'on the other hand, it is because [the mind] is at rest and has come to a halt ...' [247b11], and that could have been given in answer.[285]

> **247b13-248a2**[286] Further, just as[287] [a person passes] from being drunk or asleep {or sick and we do not say that he has come to be a knower again (even though he was incapable of using the knowledge earlier), in the same way we do not say that one becomes a knower when he originally acquires the knowledge, for that happens because the soul settles down from its natural disturbance and becomes something that has knowledge and understanding. This is why children can neither learn nor discriminate with their senses as well as adults,} for the disturbance and the motion are great.

He has proved that also the original acquisition of knowledge, that is, of being able to have the knowledge and to come to be in a state of it, does not occur through the coming to be of the knowledge, nor yet through alteration of the future knower, since [knowledge] accrues because the disturbance in the body comes to rest, and coming to rest [does] not [occur] through motion (since coming to rest is a motion and there is no motion of motion). He also proves now by means of the example of the person who is drunk or asleep or sick that it is not

because the knower is altered in any way that the change to the using and activity of the knowledge occurs, nor yet because the knower comes to be, even though the knower 'was unable to use the knowledge' [247b15-16] during the course of being asleep or drunk or sick, for the removal of what is hindering suffices. Thus, he says, it is not the case that in the person who is changing from ignorance to understanding something arises which was not present earlier, nor yet is the intellect altered, but there occurs only a halt 'of the natural disturbance' [247b17-18], i.e. a calming, and this suffices for being able to understand and to know. For on account of this disturbance, he says, 'little children too are able neither to learn nor' to perceive and 'judge' sensible things accurately 'equally' with adults [247b18-248a1]. 'For', he says, 'the disturbance and the motion are great' [248a1-2] in little children on account of nutrition and growth. Accordingly, just as little children become suited for perceiving when 1079,1
the disturbance ceases, so too the coming to rest of the disturbing [factors] suffices for learning and understanding and knowing. But those who are drunk and asleep and sick, possessing the state and hindered only with regard to the activity, are reasonably said neither to come to be knowers nor to be altered into knowers.

But if little children at some time acquire the state of knowledge and understanding after not having had it earlier, how is it that they do not come to be knowers from nonknowers? Or how is the example derived from the person who is drunk and asleep and sick similar to those acquiring the state which they did not have? Perhaps, then, if the person originally acquiring the state is similar to the person who is drunk and asleep, the soul always possesses the state, and cases of learning really are cases of recollection, as Plato's account would have it. But it is possible, as Plotinus says, 'both to have and yet not to have ready to hand',[288] and there are many varieties [i.e. levels or degrees] of what is ready to hand and not ready to hand.[289]

Little children have the state of knowledge and understanding in the sense that it is in their essence [*ousia*] though hitherto unmoved. This is why they seem to be in need of the acquisition of knowledge. We say that they learn, the learning being a kind of spur and the acquisition being already motion and preparation for activity. These, indeed, do not themselves belong to [older] children either, but [only] when the disturbance of the body is already ceasing, since earlier [the activity] is still more buried than the state and in need of much help for its manifestation.

On the other hand, the grammarian who is drunk and asleep has made the state ready[290] in some way, but [this] person who has made it ready[291] is still hindered in some way. The one who is sober and awake and healthy has the state readier to hand and more prepared, but this person too still has it held in check in some way, and although the capacity has been prepared in him, he is engaged in other things or the cognizable is not yet present, since, at any rate, the one who is active in virtue of the state has already made it completely ready for activity as well.

Aristotle began from the person who has the state ready to hand but who is not active because the cognizable is not present to the sense organ; then he introduced the person using [the state] and being active; then the person acquiring the knowledge. Perhaps he began from what is more evident, for the person who knows because the knowable has come to be becomes a knower in actuality not by having been moved at all himself, but because some other thing has come to be. And the activity in virtue of the knowledge is not an alteration of the knower, nor yet a coming to be, and, furthermore, the seeming acquisition of the knowledge does not occur because the person acquiring it is altered.

1080,1 **248a2-6**[292] [The soul] settles down and comes to rest {relative to some things under the influence of nature itself whereas relative to other things under the influence of other factors. However, in both sorts of cases it is when certain things in the body are altered, just as it is in the case of using and activity} whenever the person becomes sober and wakes up.

Having stated that when the disturbance of the hindering [factors] settles down, [people] become capable of perception and intelligent and knowing, he adds that 'relative to some things' [the soul] settles down 'under the influence of nature', that is, relative to being active with the senses more accurately, 'but relative to other things' [248a2-3] under the influence of habits and teaching, as for instance relative to the desires, though nature helps relative to them too. Perhaps relative to the natural sensory and desiderative motions they settle down under the influence of nature, but relative to the [motions] pertaining to crafts and rational disciplines under the influence of reason. But they settle down, he says, 'in both when certain things are altered' [248a3-4]. By 'in both' he might mean in those who settle down from within under the influence of nature and in those who do so from outside under the influence of habits and teaching. Or he said 'in both' [meaning] those acquiring the state originally and those being restored after drunkenness or sleep or sickness. Those settling down from within and those doing so from outside both settle down 'when certain things in the body are altered',[293] for when bodies[294] come to be wetter or drier or hotter or colder, as in the case of the person changing from being drunk to being sober, the pneuma is being restored from disproportionate to natural wetness and hotness, and for this reason [the person] is again using the knowledge who earlier was unable to use it because of the drunkenness and the bad temperament deriving from it. Accordingly, just as in the case of the latter the state is not reacquired when the person is sober, but the using is prepared for when the annoying [factor] has ceased through the alteration of 'certain things in the body' [248a4], so too in the case of the person who seems to acquire[295] the state originally.

> **248a6-9**[296] It is evident, then, from what has been stated {that alteration occurs only in sensibles (i.e. sensible qualities) and in the sensory part of the soul,} and in no other thing except incidentally.

He reminds us next of the purpose [*hou kharin*] for which all the preceding statements were made. For he proposed to prove that alteration occurs with respect to the affective qualities. These are the sensible [qualities], by which the sense organ [*aisthêsis*] is altered. Consequently, only the sensory [part] of the soul is altered in its own 1081,1
right, or rather the thing possessing the soul is altered with respect to the sensory [part].[297] However, the intellect [is] not [altered], nor are contemplating and understanding accomplished through alteration, except incidentally, for [they occur] 'when certain <of the>[298] things in the body are altered' [248a4], as has been stated in the case of the person who is sober after drunkenness; moreover, [they involve alteration incidentally] also because of being in the thing which is perceiving and being altered. But with respect to the knowledge itself there is no alteration, both because [knowledge] is a perfection and the change to a perfection seems rather to be coming to be and not alteration, and also because knowledge is among the relatives and in the relatives there is no motion in its own right.[299]

Alexander inquires: If change with respect to quality is alteration, and state and condition have been enumerated as the first species of quality in the *Categories*,[300] 'and shape and the figure connected with each thing are a fourth kind of quality',[301] how is change with respect to state and shape not alteration? And he solves [the problem] nicely on the basis of Aristotle's [own words]: for Aristotle asserts that alteration occurs only with respect to the affective qualities, and he rejects the first and fourth type of quality as not producing alteration, since those qualities which contribute to the essence [*ousia*] and the perfection [of a thing] and become the form of what partakes [of them] do not make the things changing with respect to them otherwise [*alloia*] but rather other things [*alla*].[302] States are of this kind because they are perfections and make a thing to be[303] the very thing it is said [to be], and so are shapes. This is why what has been shaped is no longer spoken of with the term for the matter, [e.g.] wood or bronze. But change with respect to hotness and coldness and dryness and wetness is not always alteration either. For example, when fire comes to be from water, the wetness and coldness having changed into dryness and hotness, such a change is not spoken of as alteration, because the change is essential [*ousiôdês*] and constitutive of form.[304] And, accordingly, changes with respect to the virtues, [changes] perfecting the underlying thing and contributing to [its] essence [*to ti ên einai*], are not alterations, nor are those with respect to shapes and figures. Rather, only [changes] with respect to the incidental qualities, which merely produce affections in the things partaking [of them] and do not invest them with essence nor constitute their form, are [alterations].

[CHAPTER 4]

1082,1 **248a10-18** One might pose the puzzle whether every motion is [comparable with every other or not]. {If indeed they are all comparable, and things moving an equal amount in an equal time are of the same speed, there will be some curved line equal to a straight one, and greater and smaller. Further, an alteration and a locomotion will be equal when in an equal time one thing has been altered and another has been moved locally. Therefore, an affection will be equal to a length, but that is impossible. Is it, then, that things are equal in speed whenever they have moved an equal amount in an equal time, but an affection is not equal to a length, so that an alteration is not equal to a locomotion, nor smaller,} so that not every [motion is] comparable?

He has concluded the original argument proving that the thing first imparting motion (not in the sense of that for the sake of which [i.e. the final cause] but rather in the sense of the source of the motion [i.e. the efficient cause]) and the thing proximately moved by it are together for every species of motion and nothing is in between [them],[305] and up to this point he has examined each species of motion by itself and given its properties. Starting here he poses as a puzzle and investigates whether every motion is comparable with every [other] with respect to the distinctive differences of motion, [namely,] faster and slower [speed],[306] so as not only to compare [motions] of the same species with one another, for instance, an alteration with an alteration (so that for him one alteration is faster than a certain alteration whereas another is slower and another is equal); moreover, whether it is possible also to compare alteration with locomotion, or with increase and decrease, so as to claim that they are equal in speed or that one is faster and the other slower.

One ought to know that 'to compare' [*sumballein*] means the same as 'to set alongside' [*paraballein*] and 'to judge together' [*sunkrinein*] and, speaking generally, 'to examine together' [*sunexetazein*].[307] The ancients used to speak of things as 'being set alongside' one another, but more recent [thinkers speak of things as] 'being judged together' [*sunkrinesthai*], whereas the ancient physicists used the term *sunkrisis* of that which brings separated things together.[308]

Having posed this puzzle, [Aristotle] proves that it is not the case that every motion is comparable with every [other] (for there is not any common measure of all [motions]), but only those of the same species [are comparable] with one another. He proves it first through [reduction to] impossibility in the case of two locomotions, that over a straight line and that in a circle, having assumed in advance what has already been proved by him, [namely,] that things moving 'an equal [amount] in an equal time' [248a12] are of the same speed,[309] and drawing the impossible [conclusion] from their being comparable that a straight line will be equal to a curved one. For if there were things which were equal in speed, one moving over some straight line and the other over a curved one (and things moving 'an equal [amount] in an equal time' are equal in speed), the straight line would be equal to the curved one.

1083,1 [In Aristotle's day,] it was still being investigated whether it is possible for a straight line to be equal to a curve, or rather it had been given up on.[310] And hence the squaring of the circle had not yet been discovered either, and, even if it seems to have been discovered now, nevertheless it is accompanied by certain disputed hypotheses. The reason why the squaring of the circle, though not yet discovered, is still being investigated, as well as [the question] whether a straight line is equal to a curved one, is that it has also not been discovered that these things are impossible, as for instance [it has that] the diagonal is incommensurable with the side. This is why the latter is not still being investigated.[311]

Having proved that the motions over these lines are noncomparable because a seeming impossibility follows for the person who hypothesizes that they are comparable, he proves next in the case of the motions of different species alteration and locomotion that in fact these are still more noncomparable. He proves this too through [reduction to] impossibility. For if anyone claims that an alteration is equal to a locomotion, that is, equal in speed, 'whenever in an equal time one thing has been altered and another has been moved locally' [248a14-15], it will result that the affection involved in the alteration is equal to the length involved in the locomotion, for things moving 'an equal [amount] in an equal time' [248a12] were [said to be] equal in speed. 'But it is impossible' [248a15] for an affection to be equal to a length, for quantity and quality are noncomparable, because equality is predicated of the one but similarity of the other.[312] Also, it is obvious that [if] it is not possible to predicate equality, it is not possible for the things moving with respect to them to be spoken of as equal in speed.[313] For how can one speak of hotness or whiteness as equal to a length?

If, then, an affection is not equal to a length, and these are not comparable with one another, the motions with respect to them will be neither equal nor comparable. For if those [motions] were [equal and comparable], these things [sc. affection and length] would be too, because things moving 'an equal [amount] in an equal time' [248a12] are equal in speed.

After stating, 'Therefore an affection will be equal to a length' [248a15], he added the cause why this is impossible.[314] For we say things are equal in speed 'whenever they move an equal [amount] in an equal time'.[315]

The 'or smaller' which is added[316] eliminates their [i.e. the motions'] being comparable at all, if [one motion is] neither equal to nor smaller [than the other], for it is obvious that it is not greater either, for where there is a greater, there is also a smaller.

Having proved that it is not the case that every locomotion is comparable with every [other] and also that alteration is not comparable with locomotion at all, he reasonably concluded that 'it is not the case that every' motion is 'comparable' with every [other] [248a18].

248a18-b6 [How will this be the result] in the case of a circle and a straight line? {It is absurd to say that this thing moving in a circle cannot be equal in speed to a thing moving in a straight line but is immediately necessarily either faster or slower, just as if one were downhill and the other uphill. Nor does it make any difference to the argument if someone asserts that it is immediately necessary for one of the things to be moving faster or slower, for then a curved line will be greater and smaller than a straight line, and so it will be equal too. It will be greater and smaller than a straight one because, if in time A one thing has traversed B and the other has traversed C, then B could be greater than C, for this is what being faster was said to mean. Therefore, since if it also moves an equal amount in less time it is faster, there will be some part of the time A in which the thing traversing the circle will go through a portion of the circle equal to the amount C moves in the whole of time A. But, in fact, if the amounts are comparable, there

> results what was just stated, namely, that a straight line is comparable with a circle.} But [a straight line and a circle are] not comparable; therefore, neither are the motions.

Having proved that [motion] in a circle is not comparable with that over a straight line, nor alteration with locomotion, he poses the puzzle why [motion] in a circle is not comparable with that over a straight line and the straight line itself with the curved one. First he asserts that it is 'absurd' [248a19] to believe that it is impossible for this thing to be moving over a straight line and that thing over a circle 'similarly' [248a20], i.e. with equal speed, but rather to think that it is 'immediately necessary' [248a21] for what is moving over one or the other of them to be moving faster, obviously taking the difference from the underlying thing [sc. from the path], as though one thing were moving downhill and the other uphill. For under such conditions it is not at all absurd for the motion even of things which by nature are equal in speed to become unequal in speed on account of the difference in the underlying things. But what difference of this kind do a circle and a straight line have relative to one another, especially if the circle is not standing upright, so that the moving thing seems to ascend and descend, but is lying flat on the ground?

Next he tries to prove that even if it is hypothesized that the motion over one or the other of them [i.e. over either the straight line or the circle] always occurs faster, whether due to the underlying lines or due to another cause, and whether the thing moving over each should be the same or different, not even this way does it result that the [motions] are noncomparable. For one of them will be greater, [namely,] the one over which the moving thing is posited to be moving faster in the equal time, whereas the other will be smaller, the one over which the slower [thing is posited to be moving]. But in things in which there is a greater and a smaller, it is possible to assume an equal too, so that there will be some curved line equal to a straight one. He stated, 'For the curved line will be greater and smaller than the straight line' [248a24-5], instead of 'For it will be greater on some occasions and smaller on others',[317] – greater if the thing moving over it is hypothesized to be moving faster, but smaller if [it is hypothesized to be moving] slower, for the faster moving thing will move a greater [amount] in an equal time, and the slower thing a smaller one.

He also proves it by setting out an illustration.[318] For if in the same time A the faster moving thing 'has traversed' the curved line B, but the slower thing [has traversed] the straight line C, it is obvious that the curve B will be greater than the straight line C,[319] since we earlier defined the thing moving a greater [amount] in an equal time as faster.[320] But, in fact, a thing moving 'an equal' [amount] 'in less' time is faster too [248b2].[321] Consequently, in some part of the time A the faster thing will move some part of the curve B which is equal to the straight line C which the slower moving thing has moved in the entire time A. Therefore, in this way too a straight line will be equal to a curve, which he has indicated by the [statement]: 'There will be some part of the [time] A in which it will go through B belonging to the circle.'[322] He takes B as the part of the circle which is equal to the [line] C. The [statement]: 'And the [moving thing] C the [line] C in [time] A as a whole'[323] has become unclear because he here called 'C' the thing moving along the straight line though not having called it [anything] earlier.

Having thus defended the argument claiming that the straight line is comparable with the curve, and the motions over them too (for each follows from the other), he omitted refuting here the fallacy of the defence [*sunêgoria*], since he is about to introduce in a more complete form the distinguishing cause of things which are comparable and those which are not; but, having reduced it to the same impossibility, he again denies that the motions and the lines are comparable, [i.e.] the straight line with the curved one. For if the lines are comparable, he says, 'there results' the impossibility 'just stated', [namely,] that 'a straight line is equal to a circle' [248b5-6]. If, then, this is impossible, a straight line and a curved one are not comparable. If these are not comparable, 'neither are the motions' over them [248b6], for they follow from one another, as the definition of things which are equal in speed also indicates, if, at any rate, things moving 'an equal [amount] in an equal time' [248a12] are equal in speed. For if the motions are comparable, the distances over which the motions [occur] are also comparable and equal; and if these [distances] which are moved in an equal time are equal, the motions are also equal in speed.

> **248b6-12** But absolutely all those things which are not homonymous[324] are comparable.[325] {For instance, why are the pencil and the wine and the musical note not comparable with respect to which is sharper? They are not comparable because they are homonymous. But the high note is comparable with the next-to-the-highest note, because 'sharp' means the same in the case of both. Is the solution, then, to say that 'fast' does not mean the same with regard to a straight line and a curved one,} and much less still in alteration and locomotion?

Having posed a puzzle on each side concerning whether or not lines, [namely,] a straight line and a curved one, and the motions over them are comparable, he wishes next in philosophical fashion to find some rule according to which it will be possible to distinguish all comparable and noncomparable motions, even if 'quickly' [*to takheôs*] and 'slowly' are predicated of all motions.

First he claims that things having the same term, [e.g.] 'fast' [*takhu*] and 'slow', not homonymously but, obviously, synonymously, are comparable.[326] Things spoken of homonymously are not comparable, as for instance the sharpness which is spoken of homonymously is not comparable with sharpness, for sharpness in flavours and that in volumes [i.e. solid shapes] and in sounds are not comparable, nor is it possible for a person making a comparative judgment to state 'whether the pencil or the vinegar or the high note' is 'sharper'.[327] But it is obvious that sharpness in sounds is comparable with that in sounds, and the high note is sharper than the next-to-the-highest note, for 'sharp' means the same in the case of both. Also, the [instances of sharpness] in volumes are similarly comparable, as well as those in flavours. 1086,1

Accordingly, fastness and slowness, even if they are predicated of both the motions, [namely,] the one over a straight line and the one over the circle, will also not be comparable on account of their being predicated homonymously, for fastness is predicated as well of motions of different species, since both alteration and locomotion are spoken of as 'fast'. But the fastness [of the one] is not for this reason comparable with the fastness [of the other], for in their cases fastness is homonymous, since also the things in whose case they are spoken of are not comparable with one another, for a quality is not comparable with [anything in] the [category of] where. Accordingly, if both the straight line and the circle, even if they have something in common to a greater degree than those do, are nevertheless noncomparable (for he assumes this as an axiom), the motions over them would also be noncomparable, on the grounds that this fastness and slowness are spoken of homonymously.

Having stated that 'the high note is comparable with the next-to-the-highest note' because ' "sharp" means the same thing in the case of both' [248b9-10], he added: 'Is it, then, that fastness is not the same here and there?' [248b10-11], as if he were saying: 'Therefore, fastness is not the same here and there', that is, in the case of motion over a straight line and that over a circle. 'And much less still', he says, is there the same fastness[328] 'in alteration and locomotion' [248b11-12], for here the difference is greater, since alteration [pertains to] quality, whereas locomotion [pertains to] the category of where, and a straight line and a curved one are both quantities and both continuous quantities and both lines. He added this on account of what will be stated a little later, [namely,] 'Some homonymous meanings are quite far removed, but others have some similarity.'[329]

One ought to know that the text of this statement is transmitted in different forms, in one place as 'But absolutely all those things which are not homonymous are comparable' [248b6-7], as Alexander also wrote, but elsewhere as 'But absolutely all those things which are not synonymous are noncomparable.'[330] Some have transferred here the text in the other seventh book, which is: 'But is it really the case that absolutely all those things which are not homonymous are comparable?'[331] It is obvious that all have the same purport.

> **248b12-15** Or, first, is it not true {that things which are not homonymous are comparable? 'Much' means the same applied to water and air, but they are not comparable. If 'much' does not, 'double' at any rate means the same thing,} for it is two to one, and [things which are double are] not comparable.

Having stated that things which are not homonymous are comparable, whereas those which are homonymous are noncomparable, he raises an objection to this by citing muchness, which seems to be predicated synonymously of water and air. For there is not, just as there is one definition [*logos*] of sharpness in flavours and another in volumes and another in sounds, one definition of muchness in water and another in air, but rather 'much' in each of them 'means the same thing' [248b13]. And nevertheless, muchness in water and in air are 'not comparable' [248b14]. He claims this because their powers do not follow [i.e. correspond to] their volumes, but rather the power of this amount of water is many times that of such an amount of air.[332] If, then, such an amount of water is much because it was able to do much with a volume of a certain size (for the essence of water lies in the power along with the volume), such an amount of air will no longer be much because it is less with respect to power, even if it should be more than the water with respect to volume. Therefore, muchness in water and in air are noncomparable. For the air which seems to be more in volume will be less in power, even though 'much' seems to be predicated synonymously of both. Or one might assert that a person ought to compare muchness in corresponding respects, setting volume alongside volume and power alongside power.

By making use of this variation, he proved something [to be] noncomparable.[333] Even though synonymy has been assumed with respect to muchness and littleness in volumes, since it is under suspicion whether 'much' is predicated synonymously of water and air on account of what has been stated, he says: But if 'much' is not predicated synonymously [cf. 248b14], one must make the argument in the case of doubleness. For in fact both in air and in water 'double' means the same thing for each, for if anything having the ratio [*logos*] which two [has] to one is double, there is the same definition [*logos*] of doubleness in the case of both. And nevertheless, water and air are

not comparable with respect to doubleness, for the air double this amount of air will not also be double such an amount of water. This again is on account of the powers in which the essence of water or air lies. For if double the air has, in addition to the volume, also double the power of half the air in volume, it is in the ratio [*logos*] of two to one; but the water which is half such an amount of air in volume is not further half in power too, and perhaps not less either, but sometimes even exceeds the power in double the air. Accordingly, doubleness in air and water cannot be comparable. For again, if one is taken to be double the other with respect to power, it will not further be double with respect to volume. Accordingly, doubleness will be something different whenever air and water are compared, not [a single thing] which is predicated of each of them.[334] And yet it 1088,1
seemed to mean the same in the case of each of them. Consequently, it is not even the case that absolutely all things which are predicated synonymously are comparable, for he wishes to add this to what has been stated in the case of things which are homonymous, in order that even if fastness is shown to be predicated synonymously of circular locomotion and rectilinear locomotion, it does not accordingly follow that these motions are in fact comparable.

248b15-249a8 Or is there the same argument [*logos*] also in the case of these {(sc. 'much' and 'double')? For, in fact, 'much' is homonymous. Further, some things even have definitions which are homonymous. For example, if one were to claim that 'much' means 'such an amount and more', the amount involved would be different. 'Equal' is homonymous, and perhaps 'one' is too, and if that is, 'two' is also. Since why else are some things comparable but others not, if indeed there were a single nature? Or is it that the primary recipient is different? A horse and a dog are comparable with respect to which is whiter because the primary recipient in both cases is the same thing, namely, surface, and likewise they are comparable with respect to magnitude. Water and voice, on the other hand, are not comparable with respect to clarity because it is present in a different primary recipient. But this way, at all events, it will obviously be possible to make all things one, saying that each is in a different recipient. Equality, sweetness, and whiteness will be the same, but each will be in something different. Further, it is not the case that a chance thing is capable of receiving any chance thing, but rather a single thing, the primary one, is capable of receiving a single thing. Well, then, is it that things which are comparable must not only not be homonymous but must also have no specific difference either in the property or in that in which it resides? I mean that colour, for example, is subject to division into species; hence, things are not comparable with respect to being coloured,} but [are comparable] with respect to white.

He has brought the two accounts [*logoi*] into collision, the one claiming that homonymous things are all not comparable and the one claiming that synonymous things are comparable, and he has proved that in the case of air and water muchness, even though seeming to be predicated synonymously of both, nevertheless is not comparable; but if that does not [mean the same thing], doubleness at any rate more evidently also has the same definition [*logos*] in both and nevertheless is not comparable on account of the variation of the volume relative to the powers.[335] Having thus made the argument [*logos*] more puzzling, he claims that even in the case of water and air the distinction propounded earlier is correct, [namely,] that homonymous things are not comparable whereas synonymous ones are comparable. For both muchness and doubleness are noncomparable in the case of water and air because in truth they too are predicated homonymously. They do not seem to be, since their definition [*logos*], which is homonymous, is also ambiguous in the same way the term is.[336] For the definition [*logos*] of muchness which says that 'much' is 'such an amount and more' [248b18] is homonymous itself too. For 'such an amount and more' in water and in air are different in species. Also, the definition [*logos*] of equals is 'being of the same measures' or '[being] of the same numbers',[337] but in their case both the term as well as the definition [*logos*] are predicated homonymously of the things of which they are predicated. This is why water and air are not comparable with respect to equality either. For it is not the case that if they are equal in their volumes they are also equal in their powers.

Also, the definition [*logos*] of doubleness which says that it is 'two to one' is homonymous. For, in fact, even 'one' itself is homonymous, since it is predicated of substance in one sense and of quality in another. But if 'one' is homonymous, 'two' is also [248b20]. A definition
1089,1 [*logos*] composed of homonymous [terms] is itself homonymous too.

In the belief, then, that he had with good reason stated that these too are predicated homonymously, he added: 'Since why [else] are some [things] comparable but others not, if indeed there were a single nature?' [248b20-1] For if doubleness were the same in the case of air and water, as it is in the case of air and air, and water and water, 'why' is air, with respect to air, double and half and comparable, and water with respect to water, but air is not further comparable with water, so that this amount of air is double this amount of water, 'if indeed there were a single nature' of doubleness in air and water, as there is in air and air, and water and water?

Having thus posed a puzzle in reply to the alleged synonymous
predication of muchness and doubleness in the case of air and water, 10
and having for this reason declared it homonymous, he adds another cause as well of their not being comparable, even if they are predicated synonymously, [namely, the fact that] the primary bodies which have received them[338] are different, for water and air are different, even if 'much' and 'double' mean the same thing in the case of each in the sense of being predicated synonymously. [He adds this] since
where the primary recipient of things predicated synonymously is the 15
same in species, those things are comparable, even if the things in which they are differ. This is obvious from the [fact that] a horse and a dog, even though differing with respect to species, are nevertheless comparable with respect to white. For, inquiring in the case of individuals, we say, ' "Which is whiter" [248b22-3], the horse or the dog?' Or we compare white lead and snow,[339] because what has
primarily received the white is the same in both, for it is surface. 20
Similarly too, a horse and a dog are comparable 'also with respect to magnitude' [248b23-4], which is predicated of them synonymously and has the same nature in each, because the primary thing in them capable of receiving magnitude is the same in species, [namely,] body. 'But water and voice are not' [248b24] comparable, neither with
respect to white[340] nor with respect to magnitude, because in the case 25
of water the surface is capable of receiving white and body magnitude, but in the case of voice there is neither surface nor body. So if, at any rate, both muchness and doubleness were not predicated of water and air primarily (each of which owes its form both to its volume and also at the same time to its power), but rather muchness or doubleness
were said either of the volume of both or of [their] power, [then water 30
and air] would be comparable [as to] whether the volume of the air or that of the water is more or greater, and double and half, and similarly in the case of the power.

But that acumen of Aristotle's, at any rate, objects damningly[341] 1090,1
to this argument too, for using this argument, he asserts, it is possible to eliminate homonymous things[342] from among the things that exist by claiming that things which are generally agreed to be homonymous are not [really] homonymous but rather mean one and the same thing; they are not to be compared [*paraballesthai*] with one another
because the same thing is in different things capable of receiving 5
them. For instance, 'white' means the same in voice and in body, but voice and body are not comparable with respect to white, because each of the things which have received it is different in species, for voice and body are different in species. Similarly too, someone will say that 'sweet' means the same in voice and in flavour, but voice and flavour

are not compared [*sunkrinesthai*] with respect to sweetness, because the things capable of receiving them are different in nature from one another. Similarly too, [one could say that] though 'equal' [means] the same in the case of water and air, water and air are not comparable with respect to equality, because things which are differing have received the same thing.

He adds another argument [*epikheirêma*] too in proof of the same [point, viz.] that one who claims that things which are the same by nature are not ever to be compared [*sunkrinesthai*] on account of being in different species is incorrect, and he asserts that a chance thing is not capable of receiving any chance thing,[343] and especially 'the primary one',[344] 'but rather a single thing' is by nature capable of receiving 'a single thing' [249a3] which is the same in species, as, for example, surface [is capable of receiving] colours and taste flavours.[345] And it is impossible either for things which are the same to exist primarily in things which differ or for things which differ [to exist primarily] in the same thing. Consequently, whenever anything exists primarily in things which differ, it is homonymous, and it will not be because water and air are different that muchness and doubleness in each of them will be different, despite being the same in nature, as the person who posed the puzzle asserts, but rather [muchness and doubleness] themselves too are different in nature and homonymous; this is why they arose in things capable of receiving them which are different in species.

To be on the safe side, he also added 'the primary one' [*to prôton*] here, when he stated, 'But rather a single thing, the primary one, [is capable of receiving] a single thing' [249a3], for some things which are capable of receiving something secondarily, though differing from one another with respect to species, can be comparable with one another with respect to that which they have received secondarily, because the thing which has received it primarily, being something single in nature itself, belongs to those differing things. For in fact a human being and a horse, though different in species, are compared [*sunkrinetai*] with respect to the white in them, because the primary thing which has received it in both is one in nature, [namely,] surface.

Having stated that 'a single thing, the primary one', is capable of receiving 'a single thing' and that for this reason things belonging to a single species are compared [*sunkrinesthai*] with one another as if they were not homonymous, he remarked that 'not being homony-
1091,1 mous' [249a4] does not suffice for being comparable, for surface is capable of receiving the colours white and black primarily, but not the colours spoken of homonymously, for of course it is not [capable of receiving] the ones in music too.[346] And nevertheless, white is not comparable with black with respect to colour, for no one claims that the one is more or less a colour than the other.[347] Rather, in addition

to not being homonymous, comparable things, being of the same genus, must not have a difference with respect to species either, but [must] be one in species too (that being indivisible) and not only in genus, and be in a thing capable of receiving them which is one in species, for this is what 'not having a difference, neither that which nor those in which' [249a4-5] signifies, for by 'that which' he means [*legei*] the species itself being compared, such as muchness or doubleness, whereas by 'those in which' [he means] that in whose case there is the comparison in that respect, such as water or air or volume or power. For things which are the same with respect to genus are thus synonymous also, but, if they have differences, they are not comparable. Which is more a colour, white or black or grey? Or which of the things having them is more coloured? However, things partaking of the colour white are compared with one another [as to] which of them is whiter, [for instance,] snow or a swan.

> **249a8-17** So too, with regard to motion, [a thing is] of the same speed {because of moving this equal amount in an equal time. If, then, in a length there has been alteration and locomotion, is this alteration therefore equal to and of the same speed as the locomotion? But that is absurd, because motion has species. Consequently, if we say that things which have been moved locally an equal length in an equal time are of the same speed, then a straight line and a curved line will be equal. Which, then, is responsible: that locomotion is a genus or that line is a genus? The time is the same, but if the lines differ in species the locomotions too differ in species.} For, in fact, locomotion has species if that over which [a thing] moves has species.

Having articulated the argument thus, and having proved that if things are going to be comparable, it is necessary [*khrê*] not only that they themselves are not homonymous, but also that there be no difference with respect to species, neither of the things with respect to which the comparison occurs (for instance, the whites)[348] nor of the things in which these are present (for instance, the surfaces being compared with respect to the whites), but both the white and the surface [must] be one in species, next he returns to the argument before us concerning motion, for the sake of which [*hou kharin*] the inquiries just described were set in motion. For he has proposed to prove that it is not the case that every motion is comparable with every [other] (for not even every [motion] of the same species is [comparable], since locomotion over a circle is not [comparable] with that over a straight line), in order that something impossible does not follow, [namely,] that a straight line is equal to a curve. Having

defined what is being sought in the case of other examples, he asserts that the situation is similar in the case of motions too. For because there are several species of motion – not only the general ones such as locomotion, alteration, increase, decrease, but also [species] with respect to each of these – it is not the case that every [motion] is comparable with every [other], but only those occurring in the same
1092,1 ultimate species which is not further also a genus.

Proceeding methodically [*hodôi*], he articulates this too. 'Things' which have moved equal lengths 'in an equal time' will be described as 'equal in speed' [249a12-13] and will be comparable, for lengths are a single species, since they are spoken of in the case of straight lines[349] and a straight line does not have several species, nor does motion over a straight line, just as the colour white does not (if it should be an ultimate species), and surface as well. If, however, during the same time the underlying thing should be single, [namely,] a straight line, but the motions different, so that with respect to half of the length (that is, of the straight line), something stretched out alongside or the half of the length itself is altered, whereas with respect to the rest something is carried along in locomotion, the alteration will no longer be stated to be equal in speed to the locomotion, even if both are motions, and even if they [occur] during the same time, because the motion over the straight line has different species, for alteration is one thing and locomotion is another.

Accordingly, even if the length over which the motion [occurs] is the same or equal, nevertheless the motion occurring over the equal straight lines in the same time must have the same species of motion too, if the moving things are going to be equal in speed. Just as motions differing in species occurring over the same underlying thing are not comparable, motions which are undifferentiated with respect to species will not be comparable either, if the things underlying them are different with respect to species. Consequently, 'things' which have both moved 'an equal length in an equal time' in virtue of locomotion would be 'equal in speed' [249a12-13] and comparable.

If, then, anyone should claim that things [moving] over a straight line and a curve are equal in speed, the straight line will be equal to the curved one. If, then, this is impossible, what is it that is responsible for an impossibility's following for the person who claims that these motions are comparable? Is it 'that locomotion' is not an ultimate species but is still a 'genus' so as to be divided into species, [namely,] that over a straight line and that over a circle? 'Or' is 'line a genus' [249a14] divided into straight and curved? For whether these motions are different in species (though both are locomotions) or whether the things over which the motions [occur] are different (though both are lines), the motions could not be comparable, for in the case of things which are comparable, the motion and the things in which the motion [occurs] would have to be of the same species.

However, [Aristotle] himself inquires 'which', therefore, is 'responsible' for the difference in the other: locomotion because it is still a genus or line because it is a genus?[350] 'For the time', being 'the same' – the same not only in species but in number – could not be responsible for this difference. But he seems to give the difference of the lines as the cause of the difference of the motions, for, having stated concerning the lines: 'But if they are other [*alla*] in species', he added 1093,1 that locomotion over a straight line and that over a circle 'differ in species in accordance with those',[351] as what comes next indicates: 'For, in fact, locomotion', he says, 'has species if that over which [a thing] moves has species' [249a16-17].

Alexander, finding it written 'but if they are in species, they also differ in species regarding those', asserts that 'of another kind' [*hetera*] is missing in 'but if they are in species'. However, in different manuscripts I have found the text thus: 'But if they [sc. the lines over which the locomotions occur] are other [*alla*] in species, those too [sc. the locomotions] differ in species.'

Alexander asserts that another text too is transmitted which is as follows: 'For the same time is always indivisible [*atomos*] in species. Or those too [sc. the motions and the lines] differ in species at the same time.' But someone has transferred that text to this location from the other seventh book, and there was no need to fuss with it here, just as there is no [need to] with the other things written there either. However, since [this text] has been presented in an unclear way, nothing prevents appending Alexander's explanation of it. He says: 'Though it has been interjected that the entire time in which the things moving simultaneously are moving is the same in number and indivisible, one must join to the [question], "Which, then, is responsible: that locomotion is a genus or that line is a genus?" [249a13-14], the [statement], "Or those too differ in species at the same time." For [the statement], "Or those too differ in species at the same time", that is, the one follows the other, replies[352] to the puzzle. [Aristotle] indicated the original difference from which the difference in the other follows by adding: "For, in fact, the locomotion has species if that [sc. the line over which a thing moves] has species" [249a16-17]. For if the underlying things are differing in species, it is necessary that the motions involving them also differ in species. But, in fact, says Alexander, it is not true conversely in all cases, for it is not the case that motions differing with respect to species always occur over underlying things which are different with respect to species. With respect to colours, at any rate, changes differing with respect to species occur in the case of an underlying thing which is of the same species, [namely,] surface. Similarly too, if the lines differ in species,

it follows that the motion over them also differs in species. But perhaps, says Alexander, in the case of locomotion it is also convertible, for, in fact, if the locomotions differ in species, the underlying things too will differ in species. Or rather, he says, there is the same argument both in the case of alteration and in the case of locomotion.
1094,1 For the role [*logos*] which the underlying things [sc. the lines] play in the case of locomotion is played in alteration by the colours with respect to which the change in the surface occurs. Just as for locomotion there is reference to the underlying things (since in the case of locomotion [there are] the straight line and the curved one), correspondingly in surface there are white and black. [Aristotle] himself too will speak, [Alexander] says, about this further on, for, in reality, just as things being altered display a difference corresponding to the difference of the colours with respect to which they are altered, so too things moving locally differ in accordance with the difference of the intervals with respect to which they move locally.'

In this passage [in the *Physics*], the following statement seems to be unclear to me [sc. Simplicius]: 'If, then, in this [part] of the length it has been altered, but the other has moved locally' [249a9-10]. For how is anything said to be altered in half, perchance, of a straight line? Unless, perhaps, the thing moving locally in half of the straight line should also be being altered at the same time it is moving locally, whereas in the other half either it itself or another thing should only be moving locally. Or if some body stretched out alongside half the length has been altered with respect to that much length but has moved locally with respect to the other half, each in an equal time. For it is not possible to state then that the alteration is of the same speed as the locomotion, even if the time is equal,[353] and even if the distances of the length with respect to which it is altered and with respect to which it moves locally are equal. And perhaps for this reason he stated that 'in this [part] of the length it is altered, but the other moves locally' [249a9-10], in order to prove that, even if someone hypothesizes that the length associated with the alteration is equal, not even in this way are the motions comparable.[354]

249a17-21 But sometimes[355] [locomotion has species] whenever that by means of which [it occurs has species]. For instance, if feet [are the means, the locomotion is] walking, {but if wings are, then it is flying. Or not, but locomotion is different because of the 'shapes' of the paths, so that things moving the same magnitude in an equal time would be equal in speed, 'the same' in this case meaning both undifferentiated in species and undifferentiated in motion.} Consequently one must consider what is a differentia of motion.

Having stated that locomotion comes to be different with respect to species whenever that over which the locomotion [occurs] is different with respect to species, he adds that also if the means by which the locomotion occurs is different with respect to species, the locomotion too will be different with respect to species. For instance, if one thing moves by means of feet and another by means of wings, there occurs a difference in the motions too, for in fact [locomotion] by means of feet is walking, but that by means of wings is flying, and yet it is possible for both to occur over a straight line.

Having stated this, he adds: 'Or not, but the locomotion is other [*allê*] because of the shapes [*skhêmata*]' [249a18-19], not wanting 1095,1
motion to become different with respect to species if the organs by means of which the motion [occurs] differ with respect to species and not in addition the things over which [the motion] occurs. For it is because of the shape alone, he asserts, that the motions flying and walking are different [*heterai*] and not because of the species [of the organs], whenever both occur over a straight line. This is why these [motions] are comparable, for in fact we do assert that a thing which is flying and a ship which is running with a fair wind are moving faster than a thing which is walking.

But how is it that flying does not differ more with respect to species from walking than walking in a straight line does from [walking] in a circle? [We may reply that] in another respect perhaps it does differ more, but it does not differ with respect to the present distinction. For, if the [locomotions] were claimed to differ with respect to species if the things over which [they occur] are different species of line, one straight and the other a circle, it is obvious that if the things over which the locomotion [occurs] are not different, the locomotion will also not be different in virtue of the difference being investigated now, in virtue of which the motions are noncomparable. For, in fact, having proposed at the beginning to investigate which motions are comparable with respect to [being] faster and slower and equal and which are not, he claimed that [motions which are] different with respect to species are not comparable and that those occurring over [lines] differing with respect to species differ with respect to species. But flying and walking do not occur over [lines] differing with respect to species, for both can occur over a straight line and both over a circle.

He concludes next what things are equal in speed and comparable in general: those 'moving the same' distance 'in an equal' time (due to locomotion, obviously) are 'equal in speed' [249a19]. 'The same magnitude'[356] does not [mean] absolutely [the same] in number, but rather [the same] in species, so that it is in species that it is 'undifferentiated'. Such is a straight thing relative to a straight thing, and a curved one relative to a curved one. However, that which is 'undifferentiated in species' in magnitude is immediately 'undifferentiated in motion too' [249a20] with respect to species.

Earlier he stated that if that over which a thing moves has species, 'the locomotion too has species' [249a16], that is, differences with respect to species, but now [he states] that if the underlying thing is undifferentiated in species then the motion is undifferentiated too, making the theory more concise. Whenever, he says, we are investigating[357] what kinds of motions are comparable and what kinds noncomparable, it is not necessary to consider the underlying thing [sc. the line], but it suffices to consider whether the motion itself whose [status as] comparable or not we are investigating is different with respect to species, since the motion is readily accessible to us. For the underlying thing too is invariably similar, and we do not at all need to look at both. Consequently, also when we were investigating which is responsible for [the locomotions'] displaying differences – whether it is 'that locomotion is a genus or that line is a genus' [249a14] – we were investigating in vain. For the underlying thing is
1096,1 not at odds with the motion over it so that the one displays differences but the other is undifferentiated. For in fact it has been stated that 'a chance thing [is] not capable of receiving any chance thing, but rather a single thing, the primary one, [is capable of receiving] a single thing'.[358] Or rather he added the [statement], 'Consequently one must consider what is a differentia of motion' [249a21], on the grounds that it is needful for one who is going to know what kind of motion is comparable and what kind noncomparable to have previously distinguished the differentiae of motions for oneself, for the differentiae of lines are obvious in advance even to laypersons.

Alexander asserts that he [sc. Aristotle] has stated several things about noncomparable motion because it has not yet been proved that a straight line is not equal to a curve but has remained being investigated.[359] This is why [Aristotle] claims universally that the difference with respect to species of the motions follows from the difference with respect to species of the lines.[360]

249a21-25 And this argument means that the genus is not some single thing. {It is because of this that many such cases escape notice. Some homonymous meanings are quite far removed from one another, others have some similarity, and others are close, either in genus or by analogy.} This is why they do not seem to be homonymous meanings though they are.

Having proved that motions of the same species whose species is not further divided into other species are comparable on account of being a single nature,[361] but that those differing with respect to species are noncomparable, he infers as though it were a corollary based on what has been stated that things of the same genus are not comparable, because they are not one in species[362] but rather in genus. 'The genus' is not 'some single' [249a22] species like the ultimate species, which is not further divided into species (for that which is divided into several species is a genus), and consequently things which are of the same genus and not of the same species are not comparable. And, he says, 'because of this many [cases] escape notice'[363] which, though they are not, are believed comparable, as, for example, a straight line and a curved one, because, taking line to be a genus and divided into these [as species] as though the genus had some single nature, [people] believe that the straight line and the curved one are comparable on the grounds that they are of like nature.

Having stated that supposing the genus to be 'some single thing' [249a22] is responsible for some people's believing that things of the same genus are comparable, he also adds the deception resulting in connection with homonymous things, on account of which, even though all homonymous things are of different natures (since they have only a term in common but a different essence [*ousia*]) and are noncomparable, nevertheless, some of them seem comparable. For among homonymous things,[364] he says, there is much difference, for in fact 'some are quite far removed',[365] like the ones due to chance, 'whereas others have some similarity',[366] as for example images relative to the originals, 'and others are near either in genus or by analogy'.[367] Things spoken of as derived from a single thing [*aph' henos*] and relative to a single thing [*pros hen*] are [near] 'in genus', for there is something common to them, [namely,] that from which [they derive] and that to which they are relative, producing the appearance of a genus.[368] Things preserving the similarity in the 1097,1
definition [*logos*] and the [linguistic] usage are homonymous 'by analogy', as for example *arkhê*, for a spring and a heart and a unit and a point and the leading [element] in a city are spoken of as an *arkhê*, though differing so much in nature from one another.[369] Taking, then, things which are near and similar to be of like nature, [people] suppose them comparable, like muchness and doubleness in the case of water and air.

It is possible that the [statement], 'And this argument means that the genus is not some single thing' [249a21-2], was made about motion. For in fact he stated earlier[370] that locomotion and alteration are not comparable, though they are motions, 'because motion has species'[371] and is not some single species or single nature, just as line is not either. Anyone believing that motion is some single indivisible species (though in truth it is a genus),[372] positing that all motions are of like nature, believes that all [motions] are comparable, even those not [generally] believed comparable.

But he [sc. Aristotle] seems to take motion as a genus in the sense that that a common feature is predicated homonymously of the different motions, because [motion] with respect to place is one thing, that with respect to quality is another, and that with respect to quantity another, since these genera too [sc. place, quality, and quantity] differ from one another. And for this reason, he adduces the difference of the homonymous meanings as responsible for the error about comparable and noncomparable motions, claiming that some homonymous motions, on account of similarity, seem to be of like nature and hence comparable.

249a25-b19 When, then, is the species of another kind? {Is it when the same thing is in a different thing or when different things are in different things? By what standard will we judge that white and sweetness are the same or different? Is it because they are in a different thing that they appear to be of another kind, or is it because they are not the same at all? How will one alteration be equal in speed with another? If becoming healthy is being altered, and it is possible for one person to be cured quickly and another slowly, and some simultaneously, then the alterations will be equal in speed, for the patients have been altered in an equal time. But altered what? We do not speak of equality here, but rather similarity plays the role with respect to quality that equality does with respect to quantity. Let things changing 'the same' in an equal time be equal in speed. Should we compare that which has the affection or the affection? Here we can assume that health is 'the same' and that it is present to an equal degree. But if the affection is different, as when one thing is turning white and another is becoming healthy, nothing is 'the same' or 'equal' or 'similar', insofar as the affections already produce different species of alteration, and there is not a single alteration, just as the locomotions were not a single one either. Consequently one must grasp how many species of alteration and of locomotion there are. If, then, the things which are moving differ in species and they are moving in their own right and not incidentally, the motions will differ in species. If the things differ in genus or in number, the motions will differ correspondingly. But should one look at whether the affection is 'the same' or 'similar' to see whether the alterations are equal in speed, or at what is being altered (for instance, if this amount of one thing is turning white and this amount of the other)? Or should we look at both, and the alterations will be the same or different according to the affection,} but equal or unequal if[373] that [sc. the thing being altered] is unequal?[374]

Having stated that things differing with respect to species are noncomparable, whereas undifferentiated things are comparable, he inquires next, 'When[375] is the species of another kind [*heteron*]' [249a25-6], i.e. different [*diaphoros*]? Is it whenever the underlying things differ and the form [*eidos*] in the different underlying things is of another kind? For instance, is the white in a horse and in a human being of another kind with respect to species on account of the difference in the underlying things, or is it not sufficient that only the underlying thing displays a difference, but in addition to the difference of the underlying thing the form [*eidos*] itself must be different too, so that things which are said to differ with respect to species [*eidos*] have 'another' form [*eidos*] 'in another' underlying thing [249a26]? For in fact it has already been stated that 'a single thing' is capable of receiving 'a single thing'[376] and that the underlying things and the forms in them follow mutually from one another. If the recipients are of another kind, the things in them are also of 1098,1
another kind, and if the underlying things are of the same species, the things in them are also of the same nature. For the primary things that surface, which is of the same species, is capable of receiving are of the same nature with one another, like colours for instance, even if they do not happen to be of the same species.

Accordingly, he inquires, 'What standard' is there? and by what 'will we judge' when 'white' or 'sweetness' means the same [249a27-8], in the sense of having one and the same nature, as for instance when white is observed in a horse and a dog, or [when it means] different things, [as for instance] when [white is observed] in a surface and in a voice,[377] and [when] the sweetness is that in flavour and in voice? Accordingly, in order that we not go astray among the particulars, what standard and criterion would there be for such a distinction? Is perception, then, as it seemed to Protagoras, an adequate criterion of such things, so that one is to believe that what appears to each person according to perception is also true, whether perception reports that it is the same or different [*heteron*]?[378] [We may reply that] one must not entrust the judgment to perception and to the appearance deriving from it, which in many cases are deceived, but rather [one must] inquire about the nature and existence of things through reason, since perception, at all events, judges the fastness and slowness of motion over a straight line and over a circle as [being] of like nature.[379]

Having posed these puzzles with locomotion in mind, even if he did not name locomotion but rather mentioned white and sweetness, he extends the inquiry to the other motions and changes as well, inquiring how we must [go about] finding the things which are comparable and noncomparable in the case of those too by distinguishing what is the same and not the same.

Mentioning alteration first, he proves that alterations too are comparable, for 'it is possible for one person to be cured quickly and another slowly'[380] and, obviously, 'some simultaneously too' [249a31]. 'If', then, 'becoming healthy' is 'being altered' [249a30],[381] 'alteration' occurring in an equal time 'will be equal in speed' [249b1]. But it is[382] not simply things moving in an equal time which are equal in speed in locomotion, but rather when they move an equal [amount] in an equal [time],[383] and distances which are of the same measures are equal.[384] But in the case of alteration, 'what' [249b2] would that be with respect to which the thing being altered is altered? Obviously quality, for change with respect to quality is spoken of as alteration. But 'equality [*to ison*] is not spoken of' in the case of quality, 'but rather, just as equality [*isotês*]' is spoken of 'in the [category of] quantity', so 'similarity' is in the [category of] quality [249b2-3], as also distinguished in the *Categories*.[385] Accordingly, having rightly noted these things, he says, 'Let changing the same in an equal time [count as] being equal in speed'[386] in the case of alteration, having
1099,1 found a general term, 'the same', belonging to both equality and similarity.

Next he poses the puzzle 'whether' one ought to seek what is 'the same' in comparable alterations in the underlying thing 'in which the affection [is]' [249b5] or in the affection itself. That is, which alterations must one state to be equal in speed – those having the same underlying things which have been altered in the equal time, or those which have changed with respect to the same affections in the equal time? He asserts that in the case of the example of health put forward the [alteration] getting healthy is equal in speed because the affection, that is, the health with respect to which the change occurs in those who are becoming healthy, is the same in species (for instance, [health] of the eyes, let us say, and not another part) or simply because it is health and not another affection.

The following must be added to the health:[387] belonging 'neither more nor less but to a similar degree'.[388] For what moving faster at one time but more slowly at another is in the case of locomotions which are not equal in speed, being altered more at one time but less at another seems to be in the case of alterations which are not equal in speed. 'Yet if the affection is' [249b7-8] not the same, but they are different, so that one thing is turning white and the other is becoming healthy, there is nothing in them with respect to which they will be compared, for there is neither anything equal, as in the case of quantities, nor similar, as in the case of qualities, nor the same, which is spoken of in a general sense in the case of both.[389]

He added the cause too, for it is that the difference of the affections with respect to which the alteration [occurs] immediately 'produces' different 'species of alteration' [249b10], even if it occurs in connection with the same underlying thing. One must aspirate *hêi* ['inasmuch as'] in 'inasmuch as these already' [249b9].[390] For it was stated in the case of locomotion as well that motions of different species are not comparable. Consequently, in the case of alteration, [motions] occurring with respect to affections differing in species are not comparable either.

If, then, the differences with respect to species of locomotion and alteration are responsible for [or causes of, *aitiai*] noncomparability, one must have distinguished for oneself 'how many' species 'of locomotion' [there are] and 'how many of alteration' [249b11], in order that we may know in what way those put forward differ from one another and in what way they do not. But while he here passes over the division of each of the motions into its indivisible species, he does offer a general method derived from the common feature of things which are different. For things which differ, as he himself has taught, differ either in genus or species or number.[391] In accordance, then, with this difference, moving things differ from one another in virtue of nothing else than the respects in which they are moving (for he spoke of these as the things 'to which the motions belong' [249b12]). 1100,1
If this [difference] should be with respect to genus, the motions too will differ from one another with respect to genus; for instance, if one thing is increasing and another is turning white, the increasing will differ in genus from turning white, for in fact the one is [change] with respect to quantity, the other with respect to quality, which indeed are different in genus of being.[392] On the other hand, if the moving things differ in [the] species with respect to which they are moving, the motions too will differ in species, as turning white [differs] from turning black.[393] But if for the moving things the difference in that with respect to which they are moving is not in species but rather only in number, these motions too will differ from one another in number. And these alone will be comparable, [namely,] the motions which are of the same species and of the same genus and differing only in number. Consequently, he has here made clearer the very thing he did not articulate in the case of locomotion: that it is by the otherness [*heterotês*] and sameness of the respects in which the motion occurs that one must judge the things which are noncomparable and the ones which are comparable.

He stated that '[the things] to which the motions belong' must differ in species in the moving things 'in their own right and not incidentally'[394] in order for 'the things moving in those respects in which they are moving to differ in species' [249b12-13] (the one 'a thing turning white', perchance, and the other 'a thing becoming healthy' [249b8], for these are the things to which the motions of whiteness and health belong) and [in order for them] to differ in species in their own right (that is, with respect to turning white and getting healthy) and not [just] incidentally[395] because one [motion], perchance, belongs to a horse and the other to a human being. For this latter is not differing in species in the respect in which they are moving but rather with respect to what they are. This is why it often seems that the motion itself differs in species because the things underlying the respects in which the motion occurs differ in species, though [in fact the motion] does not differ, because the respects in which the motion occurs do not differ.

Having thus characterized the noncomparable and comparable in the case of alterations by the difference and sameness of the affection, on the grounds that alteration occurs with respect to affection, and wishing to add the difference with respect to the underlying thing too, he again poses the question whether, if 'the alterations' are 'equal in speed' [249b16], one ought only to give as the cause the affection's being 'the same' or, rather, 'similar' [249b15] (for this is the more appropriate term belonging to quality and alteration, just as 'the same' belongs to substance),[396] or also the thing which is underlying and being altered, whenever in each of the things being altered the thing being altered is equal, as for instance [when that part] of the surface which has changed to white in an equal time is equal. Having also posed this puzzle a little bit earlier, when he said, 'Which, then, ought one to compare, that in which the affection [is] or the affection?' [249b5], and having given the affection as the cause, now he adds that 'in which the affection [is]', claiming that it is necessary to look at 'both' [249b17] – at the affection [to see] if it is the same or similar, 1101,1 and also at the underlying thing [to see] if [it is] equal,[397] and believe both comparable, but similar or dissimilar or the same[398] or of another kind in virtue of the affection, but equal or unequal in virtue of the underlying thing. Thus, an alteration will be 'the same' if it is with respect to the same affection, but 'unequal' if the things which have received the affection are unequal, and, conversely, some [other alteration] will be 'equal', but not 'the same' or not similar.[399]

249b19-26 One must consider the same thing also in the case of coming to be and perishing. {Is coming to be equal in speed if something belonging to the same indivisible species comes to be in an equal time, for instance, a human being, and is it faster if in an equal time a thing of another kind comes to be? We have to say 'of another kind' (*heteros*) because we do not have a pair of terms in the category of substance as we do in the category of quality. Or, if substance is number, we can talk about a 'greater' number and a 'smaller' one of the same species. But what they have in common is nameless, and so is each of the things which we might want to compare, in the way an affection (i.e. a quality) which is greater or exceeding is 'more'} and a quantity is 'larger'.

Having offered the causes of comparable and noncomparable changes in the case of locomotion and in the case of alteration, and having omitted increase and decrease, he passed on to coming to be and perishing. He omitted increase and decrease as well-known based on what has been stated previously, for an equal addition throughout the whole to things of the same species occurring in an equal time is an increase which is equal in speed, and a subtraction like that is a decrease which is equal in speed, since equality is spoken of also in the case of increase and decrease in the way that it is in the case of locomotion. But whereas he omitted this as clear, he does inquire in the case of coming to be and perishing how coming to be is spoken of as equal in speed with coming to be and perishing with perishing.

As though thoroughly puzzled and posing a question, he asks, Is coming to be spoken of as equal in speed 'if in an equal time' [249b20] things which are the same in species and indivisible with respect to species should come to be, 'for instance a human being, but not' [a thing which is only] the same with respect to genus, for instance, 'an animal' [249b21]? And a coming to be is 'faster' whenever 'in' the same 'equal' time what has come to be is not the same but 'of another kind' [249b21-2]. For instance, if, in case human beings are coming to be, in an equal or the same time one human being might already have come to be, and another not yet, but rather some part of the things in him might have come to be, but not yet [the human being] himself.

Having stated [that a coming to be is] faster 'if' something which is 'of another kind' and not the same should have come to be 'in an equal' time [249b22], he defends himself for saying 'of another kind' [*heteros*] and not assigning some[400] distinctive term. 'For we do not have', he says, 'any pair' [249b22] which 'otherness' [*heterotês*, 249b23] in the case of change with respect to substance signifies, in the way that, for example, in the case of [change] with respect to quality 'dissimilarity' [249b23] signifies more and less. Also, in the case of [change] with respect to quantity, 'inequality' is a general [term] but signifies two things, [namely,] larger and smaller. 'Otherness' taken in a general sense in the case of coming to be, he asserts, 1102,1
does not signify two things in this way.

Wishing to find appropriate terms also in the case of change with respect to substance, he asserts that since substance, to which coming to be belongs, is composed of numbers, one must state that a coming to be is faster and slower and in general unequal in speed whenever in the same or an equal time, a 'greater' [number] belonging to one thing and a 'smaller number of the same species' [249b24] belonging to another thing should come to be, as when, in case human beings are coming to be, in the same or an equal time a 'greater' number belonging to one comes to be, such as parts which are greater in number, and a 'smaller' [number] belonging to the other.[401] When the things have been completed, the one will have come to be faster, the other more slowly, though both being composed of equal human [*anthrôpeios*] numbers. But in the case of change with respect to affection [i.e. change in quality], the general [rubric] for the comparison [*sunkrisis*] of [changes which are] unequal in speed is named 'dissimilarity' [249b23], with each of the things [falling] under the general [rubric] being spoken of as 'more' on the one hand and 'less' on the other; and in the case of [change] with respect to quantity, 'unequal' is general, and, of the things [falling] under it, one is 'larger' and the other is 'smaller'.

He himself, though having proposed to state 'each' [249b25], was content to name the ones which exceed, [namely,] 'more' and 'larger' [249b26]. But how is it that he stated that the general [rubric] in the case of coming to be is 'nameless' [249b24], if indeed 'of another kind' [*heteron*] fits it? [We may reply] that 'of another kind' is general, for both the dissimilar and the unequal are 'of another kind'.[402]

He stated [sc. in 249b23] that substance is number, either following the Pythagoreans,[403] who claim numbers are principles of existing things:

> Hear, glorious number, father of the blessed ones, father of men,[404]

and:

> And all things resemble number,[405]

or because the comings to be and constitutings of absolutely all things are accomplished in accordance with certain definite numbers of elements and parts. This is why none of the human [*anthrôpinos*] crafts, which imitate universal creation, is able to exist[406] apart from number.

[CHAPTER 5]

249b27-250a5 Since a mover is always moving something and in something [sc. in some time] {and up to some length – there will always be some time and some length, since the mover will always both be moving the thing and have already moved it – if, then, A is the mover, B is what is being moved, C is the length, and D is the time, it follows that in an equal time A will move

> half of B double the length C [case 1]; and A will move half of B the length C in half the time D [case 2], for thus it will be proportional. Further, A will also move B} and will move it half of the [line] C in half of the time D [case 3].[407]

This[408] discussion too is germane to the one concerning the comparison [*sunkrisis*] and correlation [*parabolê*] of motions, for in fact, 1103,1
correlating the mover and the thing being moved, he here inquires further into their proportions relative to one another in the case of locomotion and alteration and increase and decrease.[409] For he inquires whether, just as a force many times a certain force which is moving something moves [a thing] many times [as large as] that, so too a fraction of a force moving something will move a fraction of the thing moved by that. For instance, whether, if a hundred persons move a ship, one [person] too will move a hundredth of the ship. He also correlates the moving force and the thing being moved relative to the time and relative to the distance of the motion; for instance, [he inquires] whether the force which has moved this thing, for instance, this ship, this much distance in this much time will move half of the ship double the distance in the same time [case 1], and whether half the force will move half of the thing moved by the whole the same distance in the same time.[410] Further, [he inquires] whether, just as the whole force moves half of the thing moved by it a certain distance in a certain time the same distance in half of the time [case 2], so too half of the force will move as a whole the thing moved by the whole [force] in a certain time in double that time.[411] For these furnish some appearance of preserving the proportion, but they are not as they appear.

First he produces the argument in the case of locomotion, and thus he will generalize the things proved in the case of this to the case of the other motions as well. He assumes in advance as evident that every mover is moving something, for it cannot be imparting motion but moving nothing; rather, just as everything moving is being moved by something, so too every mover is moving something.[412] [He] also [assumes in advance] that a mover is imparting motion 'in some' [249b27] time, for motion cannot occur without time. And further, indeed, [he assumes in advance that a mover] imparts motion 'up to some [length]' [249b27-8], for a mover with respect to locomotion imparts motion for this much length. He employed the others as evident, but he also added a demonstration of this [last] one,[413] suggesting it on the basis of the things proved earlier. For if everything moving has already moved,[414] the mover too has already imparted motion because, given that motion is cut to infinity,[415] it is not possible to assume a beginning of motion.[416] If, then, the mover has moved [something], and anything that has moved [something] has moved [it] a distance 'of some quantity',[417] every mover would already have moved [something] a distance 'of some quantity', for it cannot be that something has been moved, but not some particular amount.[418] Consequently, every mover moves something and in some time and for some distance.[419]

Having assumed these things in advance as axioms, he comes next to the proportions of the mover with respect to locomotion, the thing 1104,1 being moved by it, the time, and the distance. He takes A as the moving force, B as the weight being moved by it, and he posits C as the length, i.e. distance, of the motion which the force A has moved the weight B, and D[420] as the time in which [the motion takes place].

With these things posited, he first divides the weight and asserts that the same or an equal force moves half of the weight double the distance in the same time [case 1, 250a1-3]. Then dividing the time too, he asserts that in half of the time half of the weight will be moved by the same force the same original distance C of length [case 2, 250a3], for thus the proportion is preserved. For if something should move a ten-talent weight a distance of a stade in an hour, it will move a five-talent one two stades in an hour [case 1], but only a stade in a half-hour [case 2].[421] Then dividing the distance, with the original weight remaining [constant], he divides the time as well. For, with the same things hypothesized, the force A will move the weight B half of the distance C in half of the time D [case 3, 250a4-5]. For if some force again moves a ten-talent weight a distance of a stade in one hour, it will move the same weight half a stade in a half-hour.

These three proportions are preserved in the following way: in the same time the same force moves the smaller load a greater distance [case 1], and greater by as much as the weight is smaller. Accordingly, as half of the weight is to the whole weight, so are two stades to one, since the distance is greater by as much as the weight is smaller. And, alternately, then, as the whole weight is to a stade, so is half of the weight to double the distance.[422]

The second proportion, the one claiming that half of the weight is moved the distance of a stade in half the time [case 2], is necessary in the following way: as the whole weight is to half [the weight], so is the whole time to half [the time], and, alternately, as ten talents are to an hour, so are five to a half-hour.

Alexander has put this proportion first, having omitted, I think, the first one. As second, he put the one claiming that if the whole force has moved half of the weight a distance of a stade in half the time, in the whole time it will move it double the distance of a stade.[423] And the proportion is obvious: for as a half-hour is to an hour, so is a stade to two [stades], and, alternately, as a half-hour is to a stade, so are two stades to an hour.[424] This one, keeping the distance the same but partitioning the time, in a sense follows from the one before it; however, I do not think it is found in Aristotle.[425] 1105,1

The third proportion, the one dividing the distance while the original weight remains, claims that as the distance of a stade is to half a stade, so is an hour to a half-hour, and, alternately, as an hour is to a stade, a half-hour is to half a stade.[426] Consequently, if the whole force moved the whole weight a stade in an hour, it will move it half a stade in a half-hour.

> **250a6-12** And half the strength of A[427] will move half [the weight] {a distance equal to the original distance in a time which is equal to the original time [case 4]. For instance, let half of the original force A be E and half of the original weight B be F. Since the proportion of the strength to the weight is the same, the time and distance will be equal to the original ones. Also, if E moves F the distance C in the time D, it is necessary[428] that in an equal time E will move twice F} half of C, the length.[429]

He introduces a fourth proportion, now partitioning the force for the first time, and simultaneously the weight too, and proving that half the force will move half the weight a distance equal to the original one in the whole time in which the whole force moved the whole weight [case 4, 250a6-7], for the same proportion of the force to the weight remains, which he himself gave as the cause. For as 'the strength A is to the weight B',[430] so is the half to the half, 'so that they will move [things] an equal [distance] in an equal time' [250a9].

Then he adds a fifth [proportion] too, in which he again takes half the force, but the original weight and that time D, and he claims that it will move [it] half of the length C [case 5, 250a9-12].[431] For if the whole force moved the whole weight this much distance in some time (namely, this one), half of the force will move the same weight half of the distance in the same time. For as the whole force is to the whole distance, the half also is to the half, and alternately too, for in fact in the same time the smaller force moves [things] a smaller distance, and smaller by as much as it itself is smaller. The distance being the same, the force bears an inverse [*antikeimenos*] proportion to the time, for the smaller force moves [the same weight] the same distance in more time than the greater one does, and more by as much as it itself is smaller.

The proportion which has been put forward here, in which he proved the distance to be less in proportion to the force, he will refute with the next passage [*rhêsis*] as he proceeds, on the grounds that it is not always fixed. For there is some portion of the force moving a ship, if a hundred men are pulling it a distance of a stade – [perhaps]

the force of a single man – which is unable to pull the ship not only a hundredth part of a stade, but also the least amount whatever.

1106,1 The following text of this passage [*lexis*] [i.e. 250a9-12] is also transmitted: 'And if E moves F the [distance] C in the time D, it is not necessary for E to move double the weight F half of C in an equal time.'[432] What is claimed by this text is truer and consonant with what is added [sc. by Aristotle in 250a12-19].[433]

> **250a12-19** 'But if[434] A will move[435] B as far as the [length] C in the [time] D, {the force E, which is half the original force A, will not move the original weight B a proportional amount of the length C in the time D or in any part of it, and it may not move B at all. For it is not true that if a whole force moves something a certain length, then half the force will move it either any length or in any time whatever. For in that case a single person might move a ship,} since the strength of the ship-haulers and the length which all moved the ship are divided into the same number of parts as there are men pulling.

With the passage [*lexis*] before this one he proved that if half the strength moves half of the weight a certain distance in a certain time [case 4], the same strength will also move double the weight half of the distance in the same time [case 5].[436] With this [passage], on the other hand, he denies that half the strength can move the whole weight half of the distance in the same time. It seems, then, that the earlier statement was made only in the case of those things which are admitted to move [something] or be moved,[437] for there are some things which are so related to one another that the whole is capable of moving the whole and the half the half proportionally, and the half the whole, with [in this last case] either the distance being less or the time being greater. And it is obvious that in the case of such things it is true to state that in the time in which the whole force moves the whole weight such-and-such a distance, half of the weight will be moved the same distance by half the force [case 4]; and the same [original] weight [will be moved] by half the force in the same time half of the distance [case 5], or the whole distance in double the time. For it was not stated that it will invariably impart motion, but [in the earlier passage Aristotle meant] that, if a thing does move [something], it will move [it] in accordance with the proportion set forth; for half of the force will not invariably be capable of moving double the weight. [It would be] still more [correct to say that] not every portion of the force will move the entire weight the least distance whatever in the greatest time whatever.

Accordingly, he claims that if the whole force, for instance, A, moves a certain weight B a certain distance C in a certain time D, it does not further [follow] that half of the force A, for instance, E, will move the whole weight B in the time D, or in any portion of it, some portion of the distance C which will be in the same ratio [*logos*] to C as a whole that the whole force A is to a portion of itself, which he named F.[438] For, whereas E was posited [in 250a7] to be half of it [sc. of the force A], he does not wish to make the argument [*logos*] only in the case of the half but rather in the case of any portion whatever. This is why he did not use E, but rather F, in order that as the whole force A is to a portion F of itself of whatever amount, so the whole distance C is to a certain portion of what is in it.[439] Neither, then, he asserts, will E, which is half of the force A, move the weight B in the time D, nor in any portion of the time D, a part of the distance C which will have the same ratio [*logos*] to the distance C as a whole which the portion of the force A which is taken has to A as a whole; for a portion of A will not move B as a whole a portion of C either in the time D or in [any] portion of it, since it is possible for the portion of the force A not to move B any chance distance at all, not even in a time many times D. For if the strength moving [something] should be divided proportionally with the distance over which the motion [occurs], <so that,>[440] just as the whole [strength] has moved the whole weight the whole distance, half of the strength also moves the same weight half the distance, and a third [of the strength moves it] a third [of the distance], a ship, were it hauled this much distance by a hundred men, would also be pulled by one [man] a hundredth of the distance which the hundred ship-haulers moved [it], which indeed is seen to be impossible.

[Alexander states:] 'From the example it is obvious that he has assigned "A" and "F" to the moving force, "A" for the whole and "F" for one of any amount whatever. He employed the latter in place of "E" because "E" definitely indicated the half. He produced unclearness when he stated, "For as the part of C is to it [sc. C] as a whole, so F is to A", rather than "A" is to "F".[441] But if he took F not as a part, of whatever amount, of the force but rather as half of the weight, as earlier when he said, "And [let] F [be] half of the weight B",[442] [then,] since the force A moved the weight B the distance C in the time D, it is obvious that in the same time it will move half of the weight B, that is, F, double the distance C [case 1]. Accordingly, instead of stating the ratio [*logos*] that the portion of the distance C will have to the whole of C, or the whole of C to its double, he stated "as A is to F".'[443]

Thus Alexander, who would have it that 'F' has not been employed in a novel way here and that 'as F is to A' has not been stated inopportunely in place of 'as A is to F', thought up, I think, a rather unconvincing explanation. For [if Alexander's interpretation were correct,] Aristotle would not have stated that 'not even in any [part] of D' [250a14] will half of the force A move [the weight either the distance] C or its double. For if the force which is half of A, E, is not able to move the weight B the distance C in the time D, much more could it not move [it] double the length C. It is better, then, to understand the [phrase] 'or[444] in proportion to C as a whole' [250a14-15] as applied to part of C, as stated earlier.[445]

> **250a19-27** For this reason Zeno's argument is not true.[446] {Nothing prevents the part's not setting in motion, in any amount of time, the air which a whole medimnus would when it fell. Nor does the part, when by itself, set in motion a part of the whole amount of air as large as it would when included in the whole medimnus, for nothing exists except potentially in the whole. But, if there are two movers, and each of them moves this much weight in this much time, then also when combined the forces will move the composite thing made up of the weights} an equal length and in an equal time.[447]

He has stated that it is not the case, if the whole force moved the whole weight a certain distance in a certain time, that half of the force will accordingly also move the whole weight half, or any portion, of the distance in the same time (for it is not the case that every portion of the force moving the whole weight will be capable of moving the whole, either in any amount of time whatever or any distance whatever). For this reason he refutes as well the argument of Zeno the Eleatic, which [Zeno] posed[448] to the Sophist Protagoras: 'Why, tell me, Protagoras', he said, 'does one millet seed produce a sound when it falls, or a ten-thousandth of a millet seed?' When [Protagoras] stated that it did not produce [a sound], [Zeno] said, 'Does a medimnus[449] of millet seeds produce a sound when it falls, or not?' When [Protagoras] stated that a medimnus did make a sound, Zeno said, 'Well, then, is there not a ratio [*logos*] of a medimnus of millet seeds to one [millet seed] and to a ten-thousandth of one?' When [Protagoras] asserted that there was, Zeno said, 'Well, then, will not the ratios of the sounds to one another be the same? For as the things making the sounds are, the sounds are too. This being the case, if a medimnus of millet seed makes a sound, one millet seed and a ten-thousandth of a millet seed will make a sound too.' Zeno, then, propounded the argument in this way.[450]

Aristotle refutes [it] by claiming that it is not the case that every force is capable of moving every magnitude, not even in any amount of time whatever. Accordingly, even if a medimnus moves an amount
1109,1 of air which will produce a sound when it moves, it is not necessary that a portion of a medimnus too is able to move an amount of air

which will make a sound. But also, with regard to 'the portion' of the air or of any weight whatever which a portion of the force would move when it is with the whole force, if the portion of the force should be by itself it will not move 'so great a portion'[451] of the weight. For instance, if one [person as part] of a hundred moved a hundredth of a ship along with the others, themselves too moving their respective hundredths, it is not the case that accordingly one ship-hauler by himself will move a hundredth of the ship which has been detached, even if it seemed that 'so great a portion' of the weight was being moved by each portion of the force when it was together as a whole.[452]

[Aristotle] gave the cause of this, [namely,] that when a whole exists the parts do not exist 'in the whole' [250a24-5] actually but rather potentially. Consequently, not even then [sc. when the whole force was imparting motion] were hundredths of the force moving hundredths of the weight, for what does not actually exist cannot actually impart motion or be moved.[453] But when the force has been partitioned, he says, the proportion is not preserved, though it is preserved when it is combined. For if some two things, he says, separated from one another should 'each' of them move 'each' [of two] weight[s], themselves also separated from one another, a certain distance in a certain time, [then] 'also when combined the forces' of the two movers 'will move' the two weights combined the same distance in the same time.[454] 'For it is proportional':[455] as each of the movers is to each of the things being moved, so also are both the movers to both the things being moved, and consequently they will move the sum too just as much distance in the same time.

But it is obvious that neither the one [masc.] millet seed (or the one [fem.] millet seed)[456] nor the ten-thousandth of a millet seed will make a sound, nor will the one ship-hauler move the whole ship the least distance whatever in the greatest time whatever. On the other hand, it is worth seeking the cause why, though the proportion is preserved in the case of half, perchance, of the force and the weight, it does not proceed forever. For he is not, I think, giving the parts' existing in the whole potentially and not actually as the cause being sought, but is claiming only that, inasmuch as they exist potentially in the whole, the parts were neither imparting motion nor moving, so that it is not possible, on the basis of the continuous parts, to require as a matter of necessity that [the parts] impart motion and move also when they have been detached. But it is not yet obvious why the proportion does not hold good forever. Yet anyone who claims that it does hold good forever will meet with [a result] still more puzzling than this one, [namely,] one man's moving Mount Athos, if it happened to have been detached from the earth. For if he moves a single stone belonging to Mount Athos such-and-such a distance in this much time, why will he not move the whole too some fraction of 1110,1 the distance in many times the time? Archimedes, having constructed, [using] this proportion of the mover, the thing being moved, and the distance, the weighing instrument called the *kharistion*, made that [famous] boast: '[Give me] somewhere to stand, and I will move the earth',[457] as though the proportion proceeded forever.

One must state, then, what indeed was also stated concisely earlier,[458] that it is not the case that every force is naturally constituted to move every weight either the least distance whatever or in the greatest time whatever; nor again is every magnitude naturally constituted to make a sound, but there is some limit [*horos*], both of the least force which when detached is no longer able to move even any weight whatever any distance whatever, and also of the greatest weight being moved, which, if increased, can no longer be moved any distance whatever by any of the forces imparting motion in a corporeal way.[459] In between there are limits of forces relative to weights and distances and times, within which it is possible both to be divided and combined proportionally. For all the parts of a force of this size, down to the least relative to this [weight], can each move the whole weight which the whole [force] moved, the distance decreasing or the time increasing proportionally.

It is obvious that there are certain due proportions and disproportions relative to one another of forces having the strength to impart motion, weights naturally constituted to be moved, distances, and times. For this force is naturally constituted to move this weight this much distance in this much time, and some greater weight a smaller distance in a greater time, but it is not naturally constituted to move this weight any distance whatever in any time whatever. Because of these differences, then, some are proportional to one another, but others are not.

The same cause necessitates that the proportion of the sound to the things making the sound also proceeds up to a certain [point]. For it is not the case that every magnitude can deliver so much of a blow as also to produce a sound, but the sound vanishes at some magnitude.[460] Consequently, the magnitudes, whatever size the least is, are in proportion, but the sounds of the magnitudes are not [in proportion] any longer. But, though these things involve some speculation and worthwhile inquiry in their own right, one must proceed to what comes next.

250a28-b7 Is it, then, also this way in the case of alteration and increase? {For there is something causing increase, and something being increased, and in a certain quantity of time and by a certain quantity. It is the same way with what is causing alteration and what is being altered: a thing is altered a certain quantity in virtue of being altered more and less, and in a certain quantity of time. In double the time it is altered double the amount, and double the thing is altered in double the time. Half the thing is altered in half the time, or in half the time the whole thing is altered half the amount, or in an equal time half the thing is altered double the amount. But if the thing causing alteration or causing increase causes this much alteration or increase in this much time, it is not necessary that it alter or increase half the thing in half the time, or alter or increase the whole thing half the amount in half the time, but it may cause no alteration or increase at all, perhaps, just as in the case of moving the weight.}

Having proved in the case of locomotion that the proportions of the movers and things being moved proceed up to a certain [point], but not, however, forever, he inquires next if they will be this way in the 1111,1
case of the other motions as well, that is, 'in the case of alteration and increase' [250a28-9]. And he proves that, since the other [factors][461] are the same, it is necessary for both the due proportions and ratios[462] to proceed up to a certain [point] in the case of these too. For in fact, just as in the case of locomotion one thing is imparting motion and another is being moved, so too one thing is causing alteration and another is being altered, and one thing is causing increase and another is being increased. And it is in time, moreover, that what is causing alteration is causing alteration and what is causing increase is causing increase, and it is some 'quantity' that 'the one is increasing [the other] and the other is being increased' [250a30] and [some quantity that] the one is altering [the other]. However, the quantity of the latter is measured 'with respect to more and less' [250b1], whenever the thing causing the alteration is either larger or smaller with respect to the power[463] in virtue of which it is causing alteration.[464]

If, then, the same axioms apply in the case of these too which apply in the case of motion with respect to place, the [results] derived from the axioms will also be the same. For if B is altered by A an amount C in the time D, 'in double' the time the same thing will be altered 'double' [that amount] by the same thing capable of causing alteration, and 'double [the thing B[465] will be altered] in double' the time and 'half [will be altered] in half the time' and 'in half [the time the whole thing will be altered] half [the amount]',[466] or, if the power has increased, double the power will cause 'double' the alteration 'in an equal' time.[467] Similarly too in the case of increase, for so it was proved also in the case of motion with respect to place.

However, it is 'not' further 'necessary', if this much has undergone this alteration [caused] by this much, or has undergone this increase, that 'half too' is altered or increased 'in half [the time]' or 'in half [the time] half [the amount]' by half the power,[468] or that the whole will be altered by half [the power] in double the time. For half the power is not invariably capable of moving the whole, but rather, 'if it so happens' [250b6], the half may be of such a size as not to have the strength to cause alteration or to cause increase at all. But in those instances in which the proportion is preserved in the case of locomotion, it will also be preserved in the case of these [motions], but in those in which the proportion has been proved not to proceed in the case of that, it will also not proceed in the case of this.[469]

Let this, then, be an end of the explanation to the maximum extent possible[470] of this seventh book.

Perhaps it will be none the worse to summarize this [book] too by [reviewing] the main points.[471]

Right at the beginning, then, he proves that everything moving, whether it be animate or inanimate, is being moved by some other thing besides itself.[472] Then [he proves] that there exists the thing first imparting motion and that one thing does not move another [and
1112,1 so on] to infinity, but there exists something which imparts motion though not being moved by another. He proves it on the basis that, if the things which are imparting motion and being moved simultaneously, i.e. in the same [time], should be infinite, it results that the motion of the infinite [things] occurs in a finite time, which indeed was proved impossible in the book before this one.[473]

Then he attacks what has been stated, on the grounds that nothing prevents the motions of an infinite [number of things] from occurring simultaneously in a finite time. Rather one must prove that the motion made up of all [the motions] becomes a single, infinite [motion], for in this way an impossibility will be inferred. Accordingly, he proves that it is necessary that a thing imparting corporeal motion with respect to place proximately, and not by means of another, imparts motion while being continuous with or touching what is being moved, so that all the things are somehow one and the motion of the single thing is one. But since the impossibility followed from the hypothesis claiming that the things imparting motion and being moved are infinite, he rightly added, lest anyone should say that nothing has been proved if the impossibility followed upon a [mere] hypothesis, that the hypothesis has been assumed as a possibility. When what is possible has been posited, something impossible ought not to follow [from it]. If, then, something impossible did follow [from it], it was not possible for the things imparting motion and being moved to be infinite.

Having used the [claim that] the thing proximately imparting corporeal motion either touches or is continuous with what is being moved, and having assumed on the basis of this that the things imparting motion and being moved are somehow one and hence the motion too is one and infinite, from which the impossibility follows, he next proves this very thing in the case of the three species of motion, [namely,] that the mover and what is being moved are together so that nothing is in between them.

[He proves it] first in the case of locomotion, for, of things moving with respect to place, some are being moved by themselves, and others by other things. Those being moved by themselves, as, for example, animate things, evidently have the mover and what is being moved together (for there is not anything in between soul and body), whereas things imparting motion from outside and by force are either pushing or pulling or carrying or turning, for in fact all the other motions are reduced to these. Of the four, he classifies carrying and turning under pushing and pulling, and he proves in the case of those that nothing is in between what is imparting motion with respect to place and what is being moved, producing the demonstration based on the definition of pushing and of pulling.

Then it is proved that nothing is in between what is causing alteration and what is being altered, for, if the things causing alteration and being altered with respect to the affective qualities are bodies, and neither does a body do [anything] to a body separated [from it] nor is quality active apart from body, there could not be anything in between the things causing alteration and being altered. For even if air is in between, that is what is proximately imparting motion.

Next [it is proved] that also nothing is in between what is primarily 1113,1
causing increase and what is being increased, nor yet [between] what is causing decrease and what is being decreased, since what is added [is what] primarily causes increase and what is subtracted [is what] primarily causes decrease. These things are continuous with what is being increased and decreased, and nothing is in between things which are continuous.

Having used the [claim that] a thing being altered by the sensibles is being altered with respect to the affective qualities, he proves this on the basis that alteration occurs with respect to no other quality. For neither change with respect to shape nor that with respect to state are alterations, but rather comings to be.

Further, he also proves in another way that change with respect to state is not alteration, for the virtues and the vices are instances of due proportion and disproportion. But those are relatives, and of relatives there is no alteration, nor any motion at all or change in its own right. Consequently, there is no alteration with respect to states, though they supervene when certain things are altered. Having proved these [points] first in the case of bodily virtue and vice, he passes on to the states of the soul as well and proves that these too are relatives and that both the moral virtues and those of the intellectual part come to be when the sensory [part] is altered. Next he concludes that alteration by sensibles occurs in the sensory portion of the soul with respect to the affective qualities.

Having proved in this way that nothing is in between what is imparting motion and what is being moved with respect to locomotion and with respect to increase and decrease and with respect to alteration, he inquires next whether every motion is comparable with every [other] with regard to the distinctive differences of motion, [namely,] faster and slower [speed], and he proves that it is not the case that all are [comparable], but [only] those of the same indivisible species, which differ only in number.

He proves this with respect to each motion and change. First in the case of locomotion [he proves] that [motion] in a circle is not comparable with that over a straight line, since a straight line will be equal to a curved one, which indeed is impossible. But to an even greater degree, locomotion will not be comparable with alteration, since an affection will be equal to a length. Then he raises an objection against [the view that] locomotions over a straight line and over a curve are not comparable, nor the lines themselves. He refutes the objection by reducing it to the same impossibility, [namely,] that a straight line will be equal to a curve. Having posed a puzzle on each side concerning [whether] the lines, [i.e.] the straight line and the curve, and the motions over them, are comparable or not, he offers a general rule in accordance with which it is possible to distinguish comparable and noncomparable motions, claiming that those having fastness and slowness homonymously are noncomparable, but those [having them] not homonymously but rather synonymously are comparable.

Then he raises an objection against this rule, citing the instances of muchness and doubleness, which are spoken of synonymously in the case of water and air even though water and air are noncomparable with respect to these, because the air which seems to be more or double with respect to volume is less with respect to power. He remarks that both muchness and doubleness are predicated homonymously and not synonymously of air and of water, but they do not seem homonymous because their definitions too are composed of homonymous [terms].

Then he adds yet another distinction pertaining to things which are comparable: the primary thing which has received the things predicated synonymously must be the same in species. For thus we claim that a horse is whiter than a dog, even though it differs with respect to species, because that which has primarily received the white, [namely,] surface, is the same in species in both.

He raises an objection against this account too, claiming that it is possible in this way to get rid of all homonymous things, if the things which inhere, though being the same by nature, come to be different on account of the difference of the recipients. He also notes that it is not even possible for things which are the same by nature to exist [*huparkhein*] in different species, for a single thing belongs [*huparkhei*] to a single thing primarily, as, for instance, colour [belongs] to surface.

Then he teaches that not being homonymous does not suffice with regard to being comparable, but both the things themselves and the things in which they are must be of the same ultimate species; for white and black are not colours homonymously but synonymously, and nevertheless they are not compared [*sunkrinesthai*] as to which of them is more a colour, since they are not of the same species but only of the same genus. Consequently, he asserts, in the case of motion too, only those involving the same ultimate species, which is not further a genus too, occurring in an underlying thing which is the same with respect to species, are comparable motions. He also adds the cause why certain homonymous things seem to be comparable, just like muchness and doubleness. It is because, he asserts, though some homonymous things are quite far removed from being the same by nature (as, for example, those due to chance and in virtue of analogy), others are nearer.

Since he stated that the motions of things moving equal [amounts] in an equal time are comparable, but in the case of alteration when the comparison [*parabolê*] with respect to quality is made it is not possible to speak of equality, he inquires how one ought to talk about comparable [motions] in the case of these and claims that instead of 'equal' and 'unequal' one must take 'same' and 'of another kind', and 'similar' and 'dissimilar', so that things changing in the same way or similarly in an equal time are spoken of as being altered with equal speed. When the affection[474] is similar (for instance, turning white), but the surfaces which are turning white are equal or unequal, then it results that the alteration is similar or dissimilar in virtue of the affection, whereas it is equal or unequal in virtue of the underlying thing. 1115,1

Having thus offered detailed articulation of comparable motions in the case of locomotion and alteration, and having passed over increase and decrease as obvious from these (for an equal addition occurring throughout the whole in an equal time to things of the same species is an increase which is equal in speed, and a subtraction like that is a decrease which is equal in speed), having passed over these, then, he came to coming to be and perishing. He inquires what kind of coming to be is spoken of as equal in speed or unequal in speed to a coming to be, and he asserts that they are unequal in speed whenever in things of the same species a greater number of parts of one thing and a smaller [number of parts] of another come to be in the same equal time.

On top of these he adds another subject for speculation germane to the discussion concerning the comparison of motions, for he correlates the mover, the thing being moved, the time, and the distance, and he inquires further into their proportions relative to one another, [inquiring] whether, just as a force many times a certain force moving something moves [a thing] many times [as large as] that, so too a fraction of a force moving something will move a fraction of the thing moved by that, <for instance,>[475] whether, if a hundred persons move a ship, one [person] too will move a hundredth of the ship the same distance in the same time or the whole ship a hundredth of the distance.

He produces the argument in the case of locomotion first, having assumed in advance as axioms that everything imparting motion is moving something and [is doing so] in time and over this much [distance]. First dividing the weight, he asserts that the same force moves half of the weight double the distance in the same time [case 1]. Then dividing the time, the weight having [already] been divided, he asserts that half of the weight will be moved by the same force half of the distance in half the time.[476] Then dividing the distance, with the original weight remaining, he divides the time as well [case 3]. At that point he introduces a fourth proportion, partitioning the force and the weight, and he claims that half the force will move half of the weight the original distance in the original time [case 4]. Then he takes half of the force, but the original weight and time, and he claims that it will move [it] half of the distance [case 5]. But this proportion he immediately denies, for if any part of the force will move the whole weight any distance in any amount of time whatever, a ship, were it hauled this much distance by a hundred men, would also be pulled by one [man] in the same time a hundredth of the distance which the hundred ship-haulers moved [it], or even the same distance in a hundred times the time.

Having stated that it is not the case that half of every force is able
1116,1 to move the same weight as the whole [force], or to move any weight at all, with this he also refutes Zeno's argument about the millet seed. For if a medimnus, he says, of millet seed makes a sound when it falls, one millet seed will make a sound too, and a ten-thousandth of a millet seed. He refutes this argument, then, by claiming that it is not the case that every force is capable of moving every weight, not even in any amount of time whatever, and thus not every magnitude makes a sound either. Putting the argument succinctly, he claims that when the force has been partitioned the proportion is not preserved, since it is not the case that every force, even the least, is

able to impart motion or make a sound, though, when [the force is] combined, [the proportion] is preserved.

Having proved these things in the case of locomotion, then, he claims that in the case of the other motions as well, since the principles [*arkhai*] are the same – the mover, the thing being moved, the time, and the distance – the things resulting from these will be the same too. Consequently, in those instances in which the proportion is preserved in the case of locomotion, it will also be preserved in the case of the other motions, but in those instances in which the proportion did not proceed [forever] in the case of that, it will not proceed [forever] in the case of these either.

Notes

1. Simplicius employs the *Physics'* formal Greek title, *Phusikê akroasis*, which could be translated somewhat more literally as *Lecture Course on Physics*. Despite the use of the standard title *Physics*, it bears mentioning that the Greek word *phusikê* has a broader application than the fairly narrow English term 'physics' does. As the adjective derived from *phusis*, it can be applied to anything concerned with nature, and in fact in 1037,14 it has been translated as 'having to do with nature'. Simplicius takes the addition of *akroasis* to mean that the material has been elaborated with some exactitude for presentation to an audience (4,10-11). (References consisting simply of page and line numbers are to Diels's two-volume edition of Simplicius' commentary on the *Physics* in the *Commentaria in Aristotelem Graeca*. The full references are: *Simplicii in Aristotelis Physicorum Libros Quattuor Priores Commentaria* (= *CAG* 9), Berlin 1882, 1-800, and *Simplicii in Aristotelis Physicorum Libros Quattuor Posteriores Commentaria* (= *CAG* 10), Berlin 1895, 801-1366.)

2. Simplicius notes at the beginning of his commentary on book 6 of the *Physics* that Peripatetic practice was to designate the books of Aristotle's works according to the letters of the alphabet (923,3-4; see also 1117, 4-5 at the beginning of his commentary on book 8). Accordingly, eta, the seventh letter, was used for the seventh book. (The Greeks also used the letters of the alphabet for numbering, but since digamma continued to be used for the number 6 even after it no longer formed part of the regular alphabet, the letter eta then stood for 8 rather than 7.)

3. Simplicius again describes the difference as minor in 1051,5-6. However, the differences in wording are sometimes considerable, and there are other differences as well (for instance, in the way different motions are classified). For more information about the two versions, see n. 13 below.

4. The Greek word *problêma* can have several meanings. In the Aristotelian corpus it is used in a broad sense to refer to any subject of investigation or inquiry. In the technical language of Aristotelian dialectic, a 'problem' is a question asking whether or not something is the case (*Top*. 1.4, 101b28-33). Aristotle also applies the term to a proposition proposed for proof (see Hermann Bonitz, *Index Aristotelicus*, Berlin 1870, s.v. *problêma*, 636a10-11). Simplicius tends to employ the word in its narrower senses; here he is referring to propositions 'put forward' for proof, a natural usage in view of the word's etymological connection with *proballein* (cf. *problêthen* in 1037,7 below).

5. As Ross points out, this is primarily true of the arguments in chapter 1 of book 7. See W.D. Ross, *Aristotle's 'Physics': A Revised Text with Introduction and Commentary*, Oxford 1936 (rpr. 1979), 16. Robert Wardy argues that the overlap is greater, partly on the grounds that chapters 2-5 are presented in defence of claims made in chapter 1. See his *The Chain of Change: A Study of Aristotle's 'Physics' VII*, Cambridge 1990, 86 n. 5, 112, 120.

6. In Aristotelian usage, a logical (*logikos*) argument is one based on premises which are general and abstract rather than specifically appropriate to a particular discipline. The usual implication is that the argument or method is less rigorous and authoritative than a demonstration in the strict sense would be (in line 13 Simplicius seems to be employing *apodeixeis* ['demonstrations'] in a somewhat informal sense). The description of an argument as *logikos* can also be used to suggest that it has a dialectical character, one mark of which is arguing from reputable opinions (so-called *endoxa*). Simplicius brings out these various aspects of a 'logical' approach in such passages as 440,21-34; 476,23-9; 1301,19-24; 1305,23-5; *in Cael*. 238,8-15. In the commentary on book 8, Simplicius quotes Alexander as contrasting 'logically and dialectically' with 'in the manner of physics (*phusikôs*) and demonstratively'

(1247,9-10). Indeed, in his reply, extant (beyond Simplicius' quotations) only in Arabic, to Galen's criticisms of the argument Aristotle uses in *Phys.* 7.1, 241b44-242a49, Alexander repeatedly characterizes Aristotle's procedure in the argument as dialectical (or 'logical', depending on how one translates the Arabic terms involved). The Greek original of that reply may well be the source Simplicius is using for Alexander's judgment; see n. 34 below. Simplicius seems to share this view of the book 7 proof; in his discussion of arguments offered in book 8 he says that Aristotle there proves that everything moving is being moved by something in a way 'more in the manner of physics' (*phusikôteron*) than he does in book 7 (1219,37-38). For further discussion of what it means to argue 'logically', see G.E.L. Owen, 'Logic and metaphysics in some earlier works of Aristotle', in I. Düring and G.E.L. Owen, eds, *Aristotle and Plato in the Mid-Fourth Century* (= Studia Graeca et Latina Gothoburgensia 11), Göteborg 1960, 163-90 (also in G.E.L. Owen, *Logic, Science and Dialectic*, M. Nussbaum, ed., London 1986, 180-99); and W. Charlton, *Aristotle's 'Physics': Books I and II*, Oxford 1970, x-xii.

7. See Fritz Wehrli, ed., *Die Schule des Aristoteles*, vol. 8: *Eudemos von Rhodos*, 2nd ed., Basel-Stuttgart 1969. The present passage is included as Eudemus Fr. 109. The omission is one of the main arguments for the claim that in Eudemus' day our book 7 had not been incorporated into the *Physics*. This argument goes back to V. Rose, as noted by Diels in 'Zur Textgeschichte der Aristotelischen Physik', *Abhandlungen der Königlichen Preussischen Akademie der Wissenschaften zu Berlin, Phil.-hist. Kl. I*, 1882, 1-42, at 40-1 (also in H. Diels, *Kleine Schriften zur Geschichte der antiken Philosophie*, Walter Burkert, ed., Hildesheim 1969, 199-238). Simplicius makes frequent use of Eudemus elsewhere in his *Physics* commentary.

8. See H. Schenkl, ed., *Themistii in Aristotelis Physica Paraphrasis* (= *CAG* 5.2), Berlin 1900. The paraphrase of book 7 is found on pp. 204-8 and occupies about 5 pages out of a total of 236. In 1051,9-13 below, Simplicius notes the spot at which Themistius begins his paraphrase of book 7 (at 234a11 in chapter 2) and also remarks that Themistius did not provide continuous coverage of the rest of the book.

9. Simplicius does not say who he thinks these people were. Moreover, some of his comments about Aristotle's references to groups of books making up our *Physics* seem inconsistent with the explanation given here. In 801,13-16 he says that Aristotle and his 'associates' (*hetairoi*) were accustomed to give the three books after book 5 the title *Peri kinêseôs* ('Concerning Motion') (cf. also 4,11-16; 6,9-10; 924,6-14; 1233,30-3; 1358,7-8; *in Cael.* 226,19-23, a passage in which it is the last four books which bear this title). Also, in 924,23-925,2, discussing the relations of the books of the *Physics* to one another, Simplicius points out that book 7 makes use of results proved in book 6. All these comments suggest that Simplicius is generally prepared to take the present position of book 7 as going back to Aristotle. He is well aware of Andronicus' editorial activity (as 924,19-20, for example, shows), but it seems unlikely that his tentative suggestion here is part of a well-developed theory. On these issues see Ross, pp. 1-19, and B. Manuwald, *Das Buch H der aristotelischen 'Physik': Eine Untersuchung zur Einheit und Echtheit* (= Beiträge zur klassischen Philologie 36), Meisenheim am Glan 1971, 3-7, 15-20.

10. The Greek verb *kinein* is transitive and means 'to move (something)'. To avoid the awkwardness of supplying an object, the active forms of the transitive verb will sometimes be translated 'to impart motion'. For the intransitive sense (as in 'the leaf moves') Greek regularly uses the middle-passive forms of *kinein*, but those same forms are also used for the true passive (as in 'the leaf is moved by the wind'). The general rule adopted here, following a suggestion of David Konstan's, will be to translate the middle-passive in most cases as the intransitive active but to use the English passive when the Greek form is accompanied by reference to a mover. For instance, in 'everything moving is being moved by something', both 'moving' and 'is being moved' translate middle-passive forms.

11. In 1042,7-9, Simplicius makes a similar comment about the superiority of the proof in book 8. The reference is to chapter 4 of that book.

12. In his concluding summary of book 7, Simplicius implies that he has done this (1111,28). For similar avowals elsewhere, cf. 773,10; 1363,25-6; *in Cael.* 108,20-1.

13. Two versions of chapters 1 to 3 of book 7 have come down to us. (Simplicius' comments in 1036,4-6 suggest that in his day there were two versions of the entire book; however that may be, his remarks in 1086,23-5 and 1093,10-12 imply that he had another version of chapter 4.) In Ross's edition of the *Physics*, what is taken to be the primary version (which will here be referred to as the 'A version') of chapters 1 to 3 is printed as the main text, while the other version (here called the 'B version') appears as an appendix following book 8. In addition to what Ross says in the introduction (pp. 11-19) and notes to his edition of the *Physics*, there is detailed discussion of the differences between the two versions in Wardy and in Manuwald (in the latter see especially pp. 7-13). The key point for present purposes is that Simplicius' commentary is based on the A version. The lemmas generally follow the A version, though the word order in the opening part of this one happens to coincide with the B version. (The lemmas present special problems. They may have been added or changed at a later date, and they are also not fully reported by Diels. See L. Tarán, 'The text of Simplicius' commentary on Aristotle's *Physics*', in I. Hadot, ed., *Simplicius, sa vie, son oeuvre, sa survie* (= Peripatoi 15), Berlin-New York 1987, 246-66, at 254; idem, review of P. Moraux, *Du Ciel*, *Gnomon* 46, 1974, 121-42, at 126-7.) To facilitate reference, the line numbers from Ross's text of the A version will be used for locating the lemmas (in some cases this means a departure from Diels's lineation). Where the versions are close enough to make it practicable, the corresponding B-version line numbers will be provided in a footnote. Thus, the present passage corresponds to B 241b24-33.

14. The first part of the lemma agrees with the B version in having *anankê hupo tinos* instead of the A version's *hupo tinos anankê*.

15. It would be more grammatical, perhaps, to follow MS M and read *kineitai* in place of *kineisthai* in 1037,13. However, it is likely that Simplicius has just taken over the wording from 241b34, with slightly ungrammatical results.

16. Simplicius' claim about the thesis's importance applies to more than just the rest of book 7. The key role of the claim in what is proved in book 8 is emphasized, e.g. in 1344,12-17; 1366,15-16. As Simplicius has already noted in 1037,6-8, Aristotle proves it again in book 8 (in 1251,2-3 and 1257,23-5 Simplicius remarks that it is proved in book 7 too). Interestingly enough, when Simplicius refers to this thesis outside this part of the *Physics*, it is sometimes book 7 rather than book 8 which is mentioned (cf. 326,28-9; *in Cael.* 240,2-3; 425,30-2).

17. In 265,14-15 impulse (*hormê*) is described as an 'internal source of motion'.

18. cf. 1049,6-9 below.

19. This list anticipates the later enumeration of externally caused locomotions given by Aristotle in 243a16-17. Cf. 1046,13-14 and 1049,13-15 below.

20. In book 8, what amount to the same cases are similarly treated as unproblematic. See *Phys.* 8.4, 254b24-7 and 255a4-5 along with Simplicius' commentary *ad loc.*

21. See *Phys.* 5.1, 224a21-30. In fact, to judge from 242b38, Aristotle seems to be working here with a simple opposition between things moving in their own right (*kath' hauta*) and things moving in virtue of a part. This is well-suited to his purposes here, but it is a little bit different from the classification spelled out at the beginning of book 5, which ostensibly distinguishes as three separate classes things moving in their own right, things moving incidentally (*kata sumbebêkos*), and things moving in virtue of a part. In the present passage, Simplicius at first glosses over the difference by describing the distinction employed here as one between things moving in their own right and primarily and things moving 'in virtue of another and, in general, incidentally' and by treating things moving in virtue of a part as a subclass of things moving incidentally. Then in lines 25-6 he simply reverts to the classification given in book 5. Aristotle's own practice in this matter is somewhat variable. For instance, in *Phys.* 8.4, 254b7-12, he treats things moving in virtue of a part as a subclass of things moving incidentally, as Simplicius duly notes in his commentary on that passage (1207,2-17). One point which Simplicius brings out in there is that 'in its own right' is contrasted with 'incidentally', whereas 'primarily' is contrasted with 'in virtue of a part' (1207,9-11). These contrasted pairs become important in the analysis of the argument in 241b44-242a49 (see 1042,4-6).

22. Diels indicates a lacuna here. The material added in square brackets is presented *exempli gratia*. Cf. 1207,5-6.

23. 'Illustrative' translates *ekthetikos*. However, the 'setting out' (*ekthesis*) of such concrete examples here and in 1084,20ff. does more than simply provide an aid to comprehension; the particular case appears as representative of the general case and the conclusion is accordingly taken to apply universally. Cf. *in Cael.* 213,5-7. For the use of this procedure in Aristotle's logic to determine the validity of syllogistic figures, see W.D. Ross, *Aristotle's Prior and Posterior Analytics,* Oxford 1949, 412-13 (*ad* 49b33-50a33).

24. It is characteristic of Simplicius' treatment of this argument that he brings soul and body into the discussion as examples of mover and moved, even when Aristotle appears to be talking primarily about bodies (as evidenced by his use of geometric examples). See 1039,28-1040,9 below, where Simplicius poses a question which he answers in 1041,28-31. As will be seen, Simplicius also has a tendency to interpret what Aristotle says in book 7 in terms of the latter's discussion of self-movers in book 8. One of his goals is to interpret Aristotle's arguments in book 7, which might seem to leave little room for self-movers, in a way which is consistent with the Platonic and Neoplatonic view of the soul as self-moving.

25. This passage corresponds to B 241b33-242a15.

26. Reading *diorthôsas* in 1038,23 with MSS AM. This reading fits well with Simplicius' use of *apeuthunein* in 1037,28, and in fact he even uses *diorthoun* of correcting people's thinking in 385,10 and 500,8 (see also *in Cael.* 542,23). A 'preconception' (*prolêpsis*) can be a natural belief which when widely shared, like 'common notions' (*koinai ennoiai*), guide us toward the truth, but Simplicius also talks about 'vain' (*mataiai*) preconceptions (*in Cael.* 142,7), so they can be false and need rectification (cf. *in Cat.* 6,22-26). Diels reads *diarthrôsas* ('having articulated'), citing the occurrence of the same word in 1040,14. However, while that word is appropriate there, it does not describe what Aristotle has done here. Simplicius does use *diarthroun* when talking about articulating the content of a preconception in *in Cat.* 379,18-19, but *diorthôsas* still seems preferable here.

27. 242a36-7. Simplicius omits the *auto* found before *kineisthai* in 242a37.

28. For the full label of the type of argument, see 103,7. Stoic logic made use of a small number of valid argument forms which were considered basic (and were referred to as 'indemonstrable'). There were standardly said to be five of these modes (Simplicius uses the word *tropos* for 'mode', reserving *skhêma* for the figures of the categorical syllogism, as in 1041,8.11). There is a convenient collection of texts on Stoic argument forms in section 36 ('Arguments') of A.A. Long and D.N. Sedley, *The Hellenistic Philosophers*, 2 vols., Cambridge 1987. See also Benson Mates, *Stoic Logic*, Berkeley-Los Angeles 1953. The second mode, which Simplicius refers to far more frequently than any of the others, is schematically expressed as: If the first, then the second; but not the second; therefore, not the first. This mode, then, corresponds to the rule of inference known as modus tollens. The argument in question here may be portrayed as: 'If this thing is not being moved by something, it is not necessary for it to cease moving when something else does. It is necessary for this thing to cease moving when something else does. Therefore, this thing is being moved by something.' It is characteristic of Simplicius to make use of both Stoic and Aristotelian logic, and in 1041,7-21 he discusses alternative ways of formulating the present argument as a categorical syllogism. For a brief discussion of Stoic logic, see *in Epict.* Dübner 124,14-53.

29. Aristotle proves that everything changing (which includes everything moving) is divisible in *Phys.* 6.4, 234b10-20 (see also 235a13-b5). This proof does not really come toward the end of book 6, but Simplicius may be thinking of the proof in 6.10, 240b8-241a26 that nothing partless can move (except incidentally). Interestingly, when Simplicius refers to this result in his commentary on book 8, he says that the proof comes in book 5 (see 1233,15-16.25-26).

30. cf. 1038,2 above.

31. The use of 'another' rather than 'something' here is surprising, since Aristotle has only posited that it is being moved 'by something' (241b44; 242a36-7; 242a46), and Simplicius has quoted him to that effect in 1038,25-31. The appearance of 'another' here may reflect the presupposition that the thing which is assumed to cease moving (described as 'another thing' in 242a35) is in fact the mover. Simplicius' summary at the end of his commentary on book 7 likewise states that Aristotle has here proved that 'everything moving is being moved by some other thing besides itself' (1111,30-1).

32. In line with its etymology, the Greek word *philologos* can mean 'fond of discourse'; in

Laws 641E5 it is applied to Athens (cf. Diodorus Siculus *Bibl. Hist.* 12,53,4). Elsewhere in Plato the word denotes the 'lover of argument' as opposed to the 'hater of argument' (e.g. *Laches* 188C6, which should be taken with *Phaedo* 89D1-3). Consistently with this, Plato does not contrast the term with *philosophos* (cf. *Republic* 582E8). Starting in Hellenistic times, however, the word takes on the implication of literary scholar, and a person so described can be contrasted with a philosopher (notably in Porphyry *Vit. Plot.* 14,19-20). The use of the term here, then, might suggest that Galen is not a philosopher but is engaged in somewhat prolix logic-chopping (an anonymous reviewer has suggested some irony in the use of the superlative). However, Simplicius does not appear to be using the term to disparage Galen in the way he uses *grammatikos* against Philoponus. In 708,27 he refers to Galen as 'the wonderful (or marvellous, *thaumasios*) Galen' and in 718,13 as 'the most learned (or most polymathic, *polumathestatos*) Galen'. In the present passage, then, 'most scholarly' seems to be a reasonable rendering of *philologôtatos*.

33. The original Greek version of Galen's attack is not extant, but parts of it are included in the Arabic version of Alexander's reply to it, which likewise has not survived in Greek. The context in which Galen's criticism (which, like Alexander's answer to Galen, seems to have been expressed in somewhat acerbic terms) was delivered is unclear. There is evidence in Alexander's reply which suggests that Galen's strictures might have been contained in a letter Galen wrote to Alexander's (their mutual?) teacher Herminus. Whether or not it originally took the form of such a letter, Rescher and Marmura (on pp. 2-4 of their edition cited in the next note) identify it with the work which appears in Galen's *de libris propriis liber* as 'On the First Unmoved Mover' (*Eis to prôton kinoun akinêton*) (Kühn's ed., vol. 19, p. 47, lines 9-10). One tantalizing piece of information is that in the Arabic version Galen is said to have taken some of the material for his criticism from Chrysippus; see Rescher-Marmura, p. 36.

34. The Greek original of this work is also not extant, but for an edition and translation of the work as it appears in Arabic see N. Rescher and M. E. Marmura, eds, *The Refutation by Alexander of Aphrodisias of Galen's Treatise on the Theory of Motion*, Islamabad 1965. Rescher and Marmura also provide a full discussion of what is known about the works involved. See also S. Pinès, 'Omne quod movetur necesse est ab aliquo moveri: a refutation of Galen by Alexander of Aphrodisias and the theory of motion', *Isis* 52, 1961, 21-54. Pinès's view is that Alexander's discussion is from his lost commentary on the *Physics*, though perhaps from an excursus (see Pinès, pp. 22-3, 30). Rescher and Marmura, on the other hand, argue on pp. 60-2 of their edition that Alexander's work was a separate treatise.

35. This example does not appear in the Arabic version (though cf. pp. 45-6 of Rescher and Marmura's edition). 'Sailing through rock' (*dia petras plein*) may have been a stock philosophical example of something impossible; at least it seems to appear as such in Sextus Empiricus (*Adv. Math.* 8,297). It also crops up as an example of a particular kind of predicate in Diogenes Laertius' life of Zeno of Citium (7,64).

36. Though Diels apparently follows the other MSS in reading 'AB' here in line 3, the argument seems to require reading 'AC' with MS M and the Aldine edition.

37. See 1038,8-9 above.

38. For the soul as moving incidentally with the body though unmoved in its own right, see 1259,7-20.31-5; 1348,32-3; 1353,7-8.

39. Simplicius uses almost these same words to introduce his own view in 762,27-9.

40. Instead of mention of a 'source of being' here one would more naturally expect reference to a 'source of moving' or a 'source of being moved'. However, the Neoplatonist Simplicius makes this point elsewhere about bodies viewed as extended and having parts (e.g. *in Cael.* 95,5-6; 140,16-17; cf. *in Phys.* 1363,6-8).

41. Literally, 'conversion with substitution of the contradictory' (*antistrophê sun antithesei*). In a categorical proposition, contraposition involves switching subject and predicate and negating both. When performed on a universal affirmative proposition, contraposition preserves truth value. See also, e.g. 1334,27-31; *in Cael.* 28,14-31,6; 164,12-15.

42. Simplicius has *pauesthai* in 1040,23 instead of Aristotle's *pausasthai*, though he does have *pausasthai* in his quotation of the same material in 1038,30.

43. After minor modifications to produce a standard-form universal affirmative proposition, we can represent the logical operation involved here as transforming 'All things which are not moved by something are things which do not necessarily cease moving because some other thing is at rest' to 'All things which do necessarily cease moving because some other thing is at rest are things which are moved by something.'

44. Simplicius seems to be drawing on more than one source of inspiration in this passage. He may have in mind Aristotle's arguments in *Phys*. 8.5, 257b2-13, that a self-mover (which is what Simplicius has just described) cannot be such that the mover and what is moved are each the whole thing. Since both here and in book 8 the point is made that what is moved will always be divisible (242a39-40; 242a47-8; cf. 257a33-b1; 258a21-2), an alternative would be for what is moved (and hence for the mover coinciding with it) to be partless and indivisible. But nothing meeting that description can be a body (cf. 1334,6-9), so we must be dealing with something incorporeal. Further, since this would be self-moving, it would fit the Platonic characterization of soul (reflected in the following quotation from the *Phaedrus*). According to what Simplicius says in *in Cael*. 95,4-5, only such an incorporeal entity could be a self-mover 'in the strict sense' (*kuriôs*). Since he also refers there to 'the self-constituted' (*to authuposta-ton*), it appears that we have gone beyond Plato's views to later Neoplatonism. Cf. also *in Cael*. 104,13-14; 140,15-16; *in Epict*. Dübner 97,36-44.

45. Simplicius is quoting the description of soul in *Phaedrus* 245C8.

46. Simplicius omits the conclusion of this first-figure categorical syllogism in Barbara: 'Everything moving in its own right and primarily is being moved by something.' Simplicius suggests that this is Alexander's formalization of the argument, and it is indeed found in the Arabic version. It appears on p. 20 of Rescher and Marmura's edition. Rescher and Marmura also translate the argument as it appears in Simplicius in their note 7 on p. 52 (where, however, '1051,7' should be '1041,7'). It was Pinès who first drew attention to the match between the Arabic version and what is found in Simplicius; see n. 42 on p. 29 of his article (where '1047,8-9' should be '1041,8-9').

47. The reference here is to the second figure (*skhêma*) of the Aristotelian categorical syllogism.

48. The clause in parentheses refers to the soul.

49. Simplicius has reworked the argument to give more prominence to the divisibility of what is moving. Though not a middle term in the strict sense, divisibility is treated as playing a causal role (as the middle term should in a scientific demonstration). Bringing in divisibility is crucial to Simplicius' effort to restrict to bodies only what amounts to a denial of self-motion.

50. This is a shortened version of the premise Simplicius gives in lines 11-12 above.

51. See 1039,13-27 above.

52. Simplicius is defending the argument by saying that since the actual division does not precede the stopping of the part, the hypothesis is a legitimate one. See 1041,31-1042,6 below. He is making use of Aristotle's claim in *Phys*. 8.5 that the potential existence of parts does not prevent a thing from moving in its own right and primarily (258a27-b4; cf. 1245,35-1246,16). Pinès (pp. 29, 35-6) holds that Simplicius does not accept Aristotle's argument. However, though he may feel that the one offered in book 8 is more rigorous (cf. 1037,6-8; 1042,7-9), Simplicius does seem to think that he has vindicated the present one and that he has done so better than Alexander manages to (cf. 1040,14-15). Indeed, Alexander is not portrayed as having much confidence in it at all. In 1039,15-16 Simplicius notes Alexander's suspicions about some aspects of it. Later on, in Simplicius' commentary on book 8, Alexander is quoted as pointing out that one of the alternatives Aristotle canvasses in an argument there may pose a problem for the book 7 argument (1246,30-1247,10).

53. This is the puzzle raised in 1039,27-1040,9 about the manner in which AB is moving.

54. Simplicius' own view is that everything moving in its own right is a body (1352,34; cf. 1243,25-6; *in DA* 303,37-304,1).

55. Simplicius is once again referring to the soul. He has managed to limit the scope of Aristotle's argument by confining it to moving bodies. So construed it rules out self-motion in the things Aristotle generally talks about as 'self-movers', namely, animals, all of which have bodies. For this view of the argument see 1245,27-33; *in Cael*. 706,15-17. At the same time, Simplicius has safeguarded the Platonic and Neoplatonic view of the soul as 'self-moved substance' (as it is referred to in such passages as 268,17-18; 356,11). See Wardy, p. 97 n. 6.

56. See 1040,9-12 above.

57. See chapter 4 of book 8.

58. Citing 209,5, Diels notes the possibility that *energesteron* ('more effectively') in line 9 should be *enargesteron* ('more evidently'). Diels may be right; Simplicius does not use *energesteron* elsewhere, and he uses *enargesteron* on several occasions in contexts involving proving and showing. See, for instance, 492,20; 676,31; 1252,12; *in Cael.* 33,32; 196,33; *in Epict.* Dübner 63,52.

59. This passage corresponds to B 242a15-b19.

60. The lemma has *hupo tinos kineisthai* instead of the *kineisthai hupo tinos* found in version A of the *Physics*. (There are several differences between the lemma and the B version.)

61. The four motions are taken to stand for an infinite number.

62. *Phys.* 8.7, 260a26-261a26. See 1048,24 below.

63. *Phys.* 6.7, 238a20-b22.

64. Reading *hup' allou kinoumenou* ('by another which is moving') in lines 21-2 instead of *hup' allou kinoumenon* ('being moved by another') with the MSS. First, Simplicius is quoting 242a51, which has *hup' allou kinoumenou* (as the *Physics* does again in 242a52). (Spengel's change of *kinoumenou* to *kinoumenon* in both places seems to lead to redundancy.) Second, the MSS of Simplicius have *hup' allou kinoumenou* in line 25 (though Diels follows the Aldine edition in reading *hup' allou kinoumenon* there). Third, Simplicius seems to be referring here in lines 21-2 to a restrictive addition to the phrase *hup' allou* ('by another'), and the reading *hup' allou kinoumenou* seems to provide that. Simplicius' discussion in 1042,27-8 offers further support for this reading, as observed by Ross in his note on 242a51 (p. 669).

65. In fact, some of the arguments for a first mover in *Phys.* 8.5 come close to relying simply on the claims that there cannot be an infinite series of moved movers and that there must be a first mover in order for there to be motion (see 256a17-19.28-9). By adding the point that the movers are themselves moving, Aristotle is able to construct an argument which makes use of the movers' combined motion and of results he has proved in book 6. In 1223,25-30, it is acknowledged that the present line of argument in book 7 is separate and distinct. For a discussion of three ways of begging the question, see 1167,21-1168,9.

66. Reading *kinoumenou* in line 25 with the MSS (Diels adopts the Aldine edition's *kinoumenon*) and also omitting the '<*to*>' which Diels takes over from the Aldine.

67. Simplicius is quoting and paraphrasing 242a57.

68. Simplicius is quoting and paraphrasing 242a55-62.

69. Simplicius has made the same claim in lines 10-11 above (and does so again in 1047,3). However, he is well aware that unity of the moving thing is only a necessary and not a sufficient condition for numerical unity of the motion (cf. e.g. 1255,16-19). See the next paragraph and 1044,13-14.

70. The quoted material is from 242a67-8.

71. Strictly speaking, coming to be and perishing are change (*metabolê*) rather than motion (*kinêsis*). 'Change' is the broader term, with 'motion' being confined to change in the categories of quality, quantity, and place (cf. *Phys.* 5.1, 224b35-225b9).

72. The Greek word *katêgoria* literally means 'predicate', and hence there is at least a *prima facie* difference between *katêgoriai* and the genera mentioned. Plotinus asks whether Aristotle's division should be viewed as one into ten genera sharing the common term 'being' or into ten predicates (*Enn.* 6.1.1.16-18). It was, of course, a longstanding exegetical problem whether Aristotle's categories involved a classification of expressions, concepts, or things. Simplicius reviews the different positions adopted on this question as part of the introductory material to his commentary on the *Categories* and tries to use Neoplatonic doctrine to reconcile them (*in Cat.* 9,4-13,26).

73. Simplicius uses the plurals 'whites' and 'blacks' in this passage, but the singular is better in English.

74. 242b37. Simplicius supplies the 'if it is' (*ean êi*).

75. See 1043,17-27 above.

76. Literally, 'concerning the one motion' (*peri tês mias kinêseôs*).

77. The reference is to *Phys.* 5.4. See also *Phys.* 8.8, 261b36-262a5.

78. *Phys.* 6.7, 237b23-238a19.

79. The expression *hôs tôi logôi eipein* seems to mean 'to put it in a nutshell'; cf. the similar expression with *phanai* in Damascius *De princ.* Westerink-Combès 6,22-3 [= *Dub. et Sol.* I, Ruelle 6,2-3]. Here Simplicius is probably referring to using four letters to stand for an infinite number of motions.

80. *Phys.* 6.7, 238a20-b22.

81. This passage corresponds to B 242b20-4.

82. This passage corresponds to B 242b24-243a2.

83. The lemma has *prôton* instead of Aristotle's *prôtôs*.

84. For instance, Aristotle's God in the role of the cosmic Unmoved Mover. See *Metaph.* 12.7, especially 1072a25-7, b3-4.

85. These types of motion are taken from 243a17. See 1037,18-19; 1049,13-15.

86. 242b63-64. Simplicius reverses the order of 'infinite' and 'finite'.

87. Reading *dunaton* ('possible') in line 18 with MSS AM, though Diels reads *adunaton* ('impossible') with F and the Aldine edition. Perhaps *adunaton* can be defended, but Simplicius goes on to say that even reducing the magnitudes to subatomic size would not suffice to make the combined thing finite. Hence, the natural thing for him to be dismissing here is the *possibility* of a finite sum.

88. *Phys.* 6.7, 238a20-b22. It is proved specifically for a finite magnitude in 238a32-36 and for an infinite one in 238b13-16.

89. This hypothesis is stated in an if-clause in 242b67-8. Ross, noting that in 242b70-1 (which Simplicius quotes next) Aristotle also includes mention of the magnitude's being finite, inserts '<either finite or>' in the if-clause too. See Ross's commentary *ad loc.*

90. See 1046,23-6 above.

91. For similar characterizations of the conclusion see 1042,13-14.26-7 and 1045,5 above. Simplicius' description, apparently formulated with the arguments in book 8 in mind, could apply either to the unmoved mover or to a self-mover (cf. 1231,34-1232,9; 1223,21-1224,5; 1320,28-30, where the same line of argument yields a self-mover). Aristotle's own conclusion, explicitly drawn in 242b71-2, is noncommittal; it is that there is some first thing which is both imparting motion and moving. See Wardy, pp. 101-2, 113.

92. For discussion of the difference between direct proofs and indirect ones such as the one used here, see *An. Pr.* 1,23; 1,29; 2,14.

93. For other instances of such near-personification of the argument see 1224,18; 1257,21. (It bears mentioning, however, that the subject to be supplied for 'inferred' could also be Aristotle.)

94. 243a30. In this quotation Simplicius has inserted the *hôs* ('as') which he goes on to argue for in lines 25-6 below.

95. See *An. Pr.* 1.15, 34a5-33; cf. *Phys.* 8.5, 256b10-12. For other references to this principle in Simplicius, see 1225,27-36; 1293,10-12; *in Cael.* 322,24-5; 325,6-8. As Alexander notes (*in An. Pr.* Wallies 177,11-18), it follows from Aristotle's characterization (Alexander calls it a 'definition') of the possible as 'that which, when hypothesized, results in nothing impossible' (cf. *Metaph.* 9.3, 1047a24-6).

96. Simplicius takes the hypothesis involved here to be that there is an infinite series of things which are both imparting motion and moving. In doing so he is joined by Ross (p. 671, *ad* 242b72-243a31) and Wardy (p. 107), probably correctly. Manuwald (pp. 35-6) and Cornford in the Loeb edition of the *Physics* (Wicksteed and Cornford, vol. 2, p. 217 note b), on the other hand, take the hypothesis to be that the motions involved are either equal to or greater than one another (cf. 242b47-8).

97. Aristotle's text reads *hê gar hupothesis eilêptai endekhomenê*. Simplicius wants the reader to understand *hôs* ('as') before *endekhomenê*.

98. This passage corresponds to B 243a3-11.

99. Literally, 'primary among motions'.

100. See 242b59-61, as noted in 1046,9-11 above.

101. Simplicius refers back to this proof in 1210,24-6 and in 1227,1-3. Manuwald (p. 103) notes that in fact the argumentation in the present chapter is often inductive.

102. Literally, 'since the mover is twofold'.

103. cf. *in Epict.* Dübner 4,25-6; 10,50-1; 22,22-3.

104. Simplicius uses *entautha* ('here') in the Neoplatonic sense in which it refers to the sensible world as contrasted with the intelligible world. Sometimes the term is restricted to the sublunar world, but in the present instance it seems to include the heavenly bodies.

105. Reading the singular *to hama* in line 17 with MS M and the Aldine edition instead of the plural *ta hama* found in MSS AF. The *Physics* has the singular form, which also appears in Simplicius' quotation of Aristotle's explanation later in this line. Diels reads the plural but suggests in his critical apparatus that the singular may be correct.

106. The definition of touching as involving things 'whose extremities are together' appears in *Phys.* 5.3, 226b23; 6.1, 231a22-3. Cf. *in Phys.* 870,10-871,3; 926,31; 1243,24-5; 1357,22-3; *in Cat.* 123,31. The term 'together' (*hama*) in this context means being 'in a single first place' [226b22]. Simplicius is being careful about the meaning because in chapter 1 *hama* is used with a different meaning. There it has a temporal sense and means 'simultaneous' or 'simultaneously' (and has been translated accordingly when it appears in Simplicius' commentary). See Wardy, p. 121, and Manuwald, pp. 107-9. (For Aristotle's tendency to refer to things which are touching as 'continuous' see 1054,14-15 below.)

107. 243a34-5. Simplicius has moved Aristotle's *estin* closer to the beginning of the clause.

108. *Phys.* 8.7, 260a26-261a26. See 1042,15-17 above.

109. The match between the A and B versions is looser than usual in this passage; the two cover the types of locomotion in a somewhat different order, and there are other differences too. See Wardy, pp. 127-32, 244-5.

110. This 'itself' (*auto*) is not in Aristotle's text. However, Simplicius also has it in his quotation of the same sentence in 1051,11.

111. Proved in chapter 1 of this book (241b34-242a49).

112. cf. 1037,14-17 above.

113. In 243a18, Aristotle uses the middle-passive of the Greek verb *anagein* with the preposition *eis* to mean that the different kinds of locomotion 'are reduced to' the four kinds mentioned. Simplicius likewise uses *anagein* with *eis* in this sense in this part of his commentary. However, elsewhere he also employs *anagein* with *eis* to mean 'classify under'; that at least is the meaning in his commentary on the *Categories*, where *anagein eis* appears to be synonymous with *anagein hupo* (see, for instance, *in Cat.* 292,22-6; 338,33-5; 422,25-30). Even if some of the occurrences of *anagein eis* here could just as well be rendered 'classify under' instead of 'reduce to', the translation 'classify under' will be reserved for *anagein hupo*. Another verb which Simplicius occasionally uses in connection with classification is *hupagein*; it will be translated 'subsume under'. The Greek-English Index gives all occurrences.

114. cf. 1347,33 (quoting Alexander); 1349,29-32; *in Cael.* 596,12-23.

115. Reading *auto* in line 6 instead of *auton*, on the suggestion of an anonymous reviewer. If *auton* were retained, it would have to refer to the air.

116. *Phys.* 8.10, 266b27-267a20. Aristotle also discusses projectile motion in *Phys.* 4.8, 215a14-17 and *Cael.* 3.2, 301b17-31. Simplicius' major discussions come up in the context of his commentary on those three passages, namely, *in Phys.* 1344,21-1352,18 (where Alexander is quoted extensively); 668,10-670,3; and *in Cael.* 595,15-597,5. Throwing's importance, of course, arises in part from the released projectile's being an apparent counterexample to the touching requirement (cf. *Phys.* 8.10, 266b28-30 and what Simplicius says in 1344,21-1345,8 and 1227,5-13). According to the reciprocal replacement theory, as the projectile moves forward it displaces that part of the medium in front of it, which in turn displaces another part and so on (cf. 1350,31-1351,27). As a result of a series of such displacements the area behind the projectile is continually being filled in. The exact mechanisms are often somewhat vague, but the inrushing medium is also viewed as propelling the projectile forward. In 668,32-669,2 Simplicius, quoting *Timaeus* 59A1-4, seems to find the reciprocal replacement theory in Plato, but in 1351,28-33 he takes issue with Alexander's attributing it to him, saying that Plato does not take the inrushing medium as causing the motion but only as accompanying it (the cause for Plato, he maintains, is rather 'nonuniformity and inequality'). Simplicius takes the problem presented by throwing to be 'troublesome' (*pragmateiôdes*, 1347,38), and he even quotes the tag line from *Timaeus* 54A5 about considering anyone who provides a better solution to be a friend (1350,7-9). Concerning Aristotle's theory and the eventual development of just such a better solution, see Richard Sorabji, 'John Philoponus', in Richard Sorabji, ed., *Philoponus and the Rejection of Aristotelian Science*, London-Ithaca

NY 1987, pp. 7-9; and idem, *Matter, Space and Motion: Theories in Antiquity and Their Sequel*, London-Ithaca NY 1988, pp. 144-5, and ch. 14.

117. Reading *hosai* in line 22 with MS M and the *Physics* rather than the *hosa* which Diels takes from the other Simplicius MSS. It seems clear that *hosai* should be adopted here, though the reading in MS M may be a correction based on the *Physics*. (In 1049,2 the lemma also has *hosai*, but that is not really evidence for what originally stood here.)

118. The *epeidê gar* (literally, 'for since') in 1050,22 seems a bit awkward, and Diels's punctuation in lines 22 and 24 has been modified as a partial remedy. (What Simplicius says here may be modelled on what Alexander says in 1051,17-21, and that may have resulted in a slightly anacolouthic construction.)

119. Simplicius uses the expression *hoi peri*, with all three Presocratics as object of the preposition *peri*. The natural meaning of this phrase is 'the circle of X' or 'those associated with X', but it can also be used as a periphrasis simply for the individual himself. Accordingly, it should not be taken to imply belief in the existence of a school of followers.

120. See 1051,17-20 (from Alexander) below. For the general point, see 1266,33-6 (in Simplicius' comment on *Phys*. 8.7, 260b7-15, where Aristotle himself seems to flirt with just such a view), and cf. *Phys*. 8.9, 265b22-9. For the attribution of this view to Anaxagoras, see also *Metaph*. 1.3, 984a11-16; Simplicius *in Phys*. 27,2-28,3; 163,5-8; 178,14-15; *in Cael*. 601,6-8. On the Atomists, see *GC* 1.2, 315b6-15; Simplicius *in Phys*. 154,23-7; 235,20-2; 1120,20-1; *in Cael*. 295,22-4; 632,6-8.

121. Aristotle rejects this view in *GC* 1.2, 317a20-31, though he seems more favourably disposed to such an approach in *Phys*. 8.7, 260b7-15.

122. See *Phys*. 5.1, 225a20-b5.

123. See 1036,4-7 above.

124. The material which Simplicius quotes appears in 243b27-9 in the B version. (However, Simplicius, along with many *Physics* MSS, has *kai hê spathêsis de*, whereas Ross prints *dê* instead of *de*.) The B version, then, treats tamping the woof (*spathêsis*) as a species of aggregating (*sunkrisis*) and parting the warp (*kerkisis*) as a species of segregating (*diakrisis*). On the other hand, the A version classifies tamping the woof as pushing together (*sunôsis*) and parting the warp as pushing apart (*diôsis*) (243b6-7). However, the A version may in fact suggest that tamping the woof is viewed as aggregating and parting the warp as segregating, since immediately after discussing them it goes on to refer to 'the *other* aggregatings and segregatings' (243b8; emphasis added). Manuwald (pp. 9-10) argues that the version Alexander refers to is not in fact our B version and that Alexander may not even have had access to that version.

125. As noted above in connection with the lemma (1049,1), this *auto* ('itself') does not appear in Aristotle's text.

126. Themistius' paraphrase of *Physics* 7 starts at 204,2 Schenkl (*CAG* 5.2). For Themistius' failure to cover the whole of book 7, see 1036,15-17 above.

127. The corresponding text in B is 243b25-9, but once again the match is looser than usual. Indeed, as Simplicius remarks in 1052,20-2, the B version contradicts the A version by saying that all locomotion is aggregating and segregating (see B 243b29). In 1052,24-8 Simplicius observes that it is possible to harmonize the two accounts if Aristotle is taken as maintaining in the A version not that aggregating is a subclass of pushing and segregating of pulling but that the members of each pair are really the same.

128. This word is used here as a general label for the Presocratics, with particular reference to those who propounded doctrines concerning the natural world.

129. On this view (and Aristotle's usual rejection of it), see 1050,22-1051,5 above.

130. They are 'parts' (*merê*) in the sense that the species is part of the genus. Cf. *Metaph*. 5.25, 1023b18-19.24-5.

131. Aristotle has 'or'.

132. Ross traces Alexander's second interpretation in part to a mistaken reading of *hê* ('the') as *ê* ('than') in 243b11 in the A version. (In the absence of breathing marks the two forms would have been orthographically identical.) As a result, according to Ross, 243b10-11 was taken to mean: 'At the same time it is evident that there is also no genus of motion other than aggregating and segregating.' See Ross's commentary *ad loc.* For further discussion, see Wardy, pp. 127-32, and Manuwald, pp. 112-15, 138-40.

133. The closing quotation mark is missing in Diels's text, but judging from 1052,4-5 it seems to go here.

134. The corresponding passages in the B version are 243a28-b23 and 243b29-244a21. In addition, the treatment of pulling in A 244a8-11 is found in B 243b23-4.

135. 244a2. Simplicius has *ek te helxeôs kai ôseôs* in place of Aristotle's *ex helxeôs te kai ôseôs*.

136. The feminine *mulê* here in line 26 seems to refer to the millstone in a handmill, whereas the masculine *mulos* in line 21 is more naturally taken to refer to something larger.

137. Simplicius makes this point elsewhere in 49,2 and 1167,36-7.

138. These 'continuous things' are the puller and what is being pulled.

139. cf. 1060,7-8 below. Continuity involves a higher degree of unity than touching does. Things are touching when their extremities are together (see 1048,17-18 above); things are continuous when their extremities are one (*Phys.* 5.3, 227a10-15; 6.1, 231a21-3; cf. *in Phys.* 1306,38-1307,1; *in Cael.* 46,20; *in Cat.* 123,30-2). Simplicius contrasts the two notions in 1210,24-6 and 1223,28-9. In 1210,16-17 he says that referring to things which are touching as continuous involves misuse of the term 'continuous'.

140. Aristotle's doctrine is that the elements have a tendency to move toward their natural (here called 'similar') places. For instance, to anticipate the examples about to be mentioned, in the case of earth, this natural place is the centre of the universe, and in the case of fire it is the periphery of the heavens.

141. This is B 243b23-4. The text Simplicius gives differs slightly from the one Ross prints. One could also omit the phrase *mê khôrizomenê* ('not being separate') and translate the last part of the sentence as: 'is faster than that of what is being pulled.' For a possible explanation of the origin of the phrase *mê khôrizomenê*, see Ross's note on B 243b24 (p. 730 of his commentary).

142. The A-version passage from 244a11 to *haptomenon* in 244b1 has nothing corresponding to it in the B version, but B 244a24-5 corresponds to A 244b1-2.

143. The lemma reverses the order of *kinountos* ('what is imparting motion') and *kinoumenou* ('what is being moved') as found in 244b1-2. However, when Simplicius quotes Aristotle's words in 1056,13-14, the order is correct.

144. An anonymous reviewer has suggested that the subject of 'claims' is Alexander, comparing line 25 below.

145. The words in angle brackets are added by Diels following the 'Emendator Ambrosianus'. On this person see the introductory material in the first volume of Simplicius' *Physics* commentary (*CAG*, vol. 9), pp. viii, xv.

146. As the text stands, Alexander is asking how such things as magnets and amber, which appear to pull things without contact, can be said to pull in virtue of touching. However, there may be a textual problem here. According to what Simplicius says in lines 25-6, the explanation Alexander goes on to give includes the concession, 'even if they themselves do not touch them'. Perhaps, then, we should insert *mê* ('not') before *haptomena* in line 26. This part of the sentence would then read: 'Alexander well inquires how things which pull naturally like Heraclean stone and such, though not touching the things being pulled, pull them ...'.

147. The material quoted here seems likely to be from Alexander's commentary on the *Physics* (so R.W. Sharples in his 'The School of Alexander?' in R. Sorabji, ed., *Aristotle Transformed: The Ancient Commentators and Their Influence*, London-Ithaca NY 1990, p. 99). Though no names are given in this passage, the talk about entanglement sounds Democritean, and in fact the quotation from Alexander in 1056,1-3 is included in the testimonia concerning Democritus in H. Diels, ed., *Die Fragmente der Vorsokratiker*, rev. by W. Kranz, 3 vols., 6th ed., Berlin 1951-52, vol. 2, p. 129. In the *Quaestiones* (also known as *Aporiai kai Luseis*) attributed to Alexander (edited by I. Bruns in *Commentaria in Aristotelem Graeca*, suppl. 2.2), there is a discussion of magnetism, including the theories of Empedocles and Democritus, both of whom employed effluences (*aporrhoiai*) in their explanations of it. There is no explicit mention of entanglement in the treatment of Democritus there, but some similar mechanisms may be at work (*Quaest.* 2.23, Bruns 72,10-74,30, where 72,10-28 is devoted to Empedocles and 72,28-73,11 to Democritus). See further Sharples' discussion on pp. 99-100 of the article cited above. There is a very full discussion of magnetism, including these texts and others, in A. Radl, *Der Magnetstein in der Antike: Quellen und Zusammenhänge* (= Boethius: Texte und Abhandlungen zur Geschichte der exacten Wissenschaften 19), Stuttgart 1988. The passages in Simplicius dealing with magnetism are translated and discussed on pages 117-24 of Radl's book (and the material from the *Quaestiones* appears on pp. 78-87). Simplicius, who drops the issue of magnetism here, hardly does justice to the problems raised by magnetic attraction. As Radl notes (p. 124), he is not really interested in magnetism for its own sake; his main concern is removing any threat which it might pose to Aristotle's position. There is another discussion of magnetism in Simplicius' commentary on book 8 (1345,14-1346,10), but in the illustration he uses in that passage he takes the magnet to be touching something (which in turn transmits the force to the ultimate recipient). There too Simplicius is concerned with some other issue, in that case projectile motion.

148. For this as a feature of pushing (*ôsis*), see 244a7-8. Pushing away (*apôsis*) is said to be 'away from itself' (but without the addition of 'toward another') in 243a19-20, and in 243b4-5 it is said to be 'either away from itself or away from another'.

149. 244b1-2. Here Simplicius quotes Aristotle with the correct word order. See 1055,4-5 above.

150. David Konstan has suggested in private communication that the words *baru kouphon* ('heavy light') in lines 17-18 may be meant as examples or may even be an intrusive gloss connected with line 26 or 28.

151. A bit oddly, *haper* ('which indeed') seems to change its reference from the things involved in the processes to the processes themselves. Perhaps we should read *hêper* here (cf. [Alex.] *Quaest.* Bruns 50,14). In any case, the sense is clear enough.

152. Alexander seems to take the additional passage to be concerned with coming to be and perishing, that is, with substantial change rather than with motion (*kinêsis*) in the narrow sense in which it is confined to change in quality, quantity, and place. The elements are a special case in which what would ordinarily count as alteration (e.g. changing from being hot to being cold) can constitute coming to be and perishing (cf. *in Cael.* 114,22-8). What Simplicius goes on to say suggests that he takes the alleged addition to be dealing exclusively with alteration, though in 1057,9-10 he leaves room for Alexander's view.

153. *Phys.* 8.4, 255b35-256a1.

154. This passage corresponds to B 244a25-245a26.

155. Simplicius is referring to the text Alexander found in some MSS of the *Physics*. See 1056,15-1057,6.

156. Aristotle has referred to the proximate mover as the *prôton kinoun* both at the very beginning of chapter 2 (243a32, as noted by Simplicius in 1048,15-16) and also in 243a14. Aristotle also uses the adverb *prôtôs* in referring to the action of the proximate mover in chapter 1 (242b59; cf. 1046,9-10). However, in chapter 2 Aristotle starts applying the description *eskhaton* ('last') to the proximate mover in some places (244b4 and 245a4) while still using *prôton* in others (245a8 and 245a12-13). At the end of chapter 2, he describes the proximate mover as *prôtou kai eskhatou* ('first and last'), the *kai* perhaps being epexegetic and equivalent to 'i.e.'. Cf. 1060,18-21; 1061,18-22 below. Perhaps the reason for the terminological vacillation is that the *prôton kinoun* which figures so prominently in chapter 1 is not the proximate mover but rather the first member of the series of movers (242a53, 242b72).

157. For the contrast between something's moving 'in its own right' (*kath' hauto*), incidentally, and in virtue of a part, see 1037,20-7 above and *Phys.* 5.1, 224a21-30.

158. 244b5-5*b*. It appears that Simplicius is quoting at least the opening words (*hupokeitai men gar hêmin*) of this sentence from Aristotle; he uses *phêsi* ('he says') immediately after them and then goes on to explain Aristotle's use of *hupokeitai* ('it is hypothesized'). However, those words do not in fact appear in our MSS of the *Physics*. Ross (following the lead of Prantl) takes this entire sentence to be a quotation from Aristotle and incorporates it into his text of the *Physics*. On this and further additions made to the *Physics* in this passage, see Ross's commentary, p. 673, and Wardy, pp. 139-40.

159. See *Cat.* 8, 9a28-10a10, where, however, the affective qualities are the third kind of quality.

160. See *Cat.* 8, 9a35-b9. Simplicius, as often, pluralizes *thermotês* and the other abstract nouns which appear here. This indicates that he is thinking of concrete instances. On this feature of Greek see H.W. Smyth, *Greek Grammar*, rev. by G.M. Messing, Cambridge Mass. 1956, sect. 1000; R. Kühner and B. Gerth, *Ausführliche Grammatik der griechischen Sprache*, II, 2 vols., 3rd ed., Hannover 1898-1904, sect. 348(3)(c) (vol. 1, p. 17).

161. See *Cat.* 8, 9b9-19.

162. See *in Cat.* 254,3-9. To count as an affective quality, a property must: (1) be a quality and (2) involve affection either (a) by causing a *pathos* in the perceiver or (b) by arising from a *pathos* in the possessor. Hotness as a property of a warm rock would thus be an affective quality: it meets requirement (1) and would satisfy (2a). The hotness of fire, on the other hand, is not an affective quality because, no matter how it fares relative to (2a) and (2b), it fails to meet (1) inasmuch as it is not a quality; rather, it is 'part of essence [*ousia*]' (*in Cat.* 254,8). See 1081,21-6 below.

163. The 'sensibles' (*aisthêta*) are the sensible features of bodies.

164. See *in Cael.* 607,32-608,2; cf. 611,27-8.

165. That is, on this first interpretation the term *hupokeimenê* in 244b6 is taken as equivalent to the passive participle of *hupotithenai*, 'to hypothesize'. *Keisthai* and its compounds regularly supply some of the passive forms of *tithenai* and its compounds. However, the alternative interpretation which Simplicius goes on to offer next is probably the correct one here.

166. On this second interpretation, *hupokeimenê* would be taken to mean 'underlying'; that is, Aristotle would be treating the sensible quality as the underlying substrate for such differences as variations in degree.

167. Aristotle distinguishes the two kinds of alteration in *DA* 2.5, 417a9-20, b2-19.

168. Simplicius has *hê kat' energeian ... aisthêsis* whereas Aristotle has *hê ... aisthêsis hê kat' energeian*.

169. cf. 1073,17-20 below.

170. Aristotle does not say this in the *de Sensu*. Even in the *de Anima* he talks of *receiving* (*dekhesthai*) the form without the matter, which is ambiguous between a physical reception and a mental reception. Ammonius' pupils Simplicius and Philoponus, however, take the mental interpretation, Philoponus insisting that the reception is *gnôstikôs* (cognitive) and Simplicius here describing it as an apprehension (*antilêpsis*, 1059,7). For the modern controversy on the same issue, see the chapters by Myles Burnyeat and Richard Sorabji in M.C. Nussbaum and A.O. Rorty, *Essays on Aristotle's De Anima*, Oxford 1992.

171. Following Diels's suggestion and reading *nou* ('intellect') in line 8 instead of the *nun* ('now') of the MSS. Diels mentions *noein* ('thinking') as an alternative possibility, but *nou* represents a smaller change.

172. On this feature of intellection, see Simplicius(?) *in DA* 45,22-46,3; 228,35-9.

173. See 1061,4 below.

174. 245a3-5. Simplicius has abbreviated Aristotle's somewhat fuller wording, which goes: 'In absolutely all these, at any rate, it is evident ...'.

175. See 1054,14-15 above.

176. This is a quotation from *DA* 2.7, 418a31-b1.

177. Aristotle refers to it as the 'last thing causing alteration' in 244b4 and 245a4. See 1057,13-16 above.

178. The role of the medium in both hearing and smell is discussed in *DA* 2.7, 419a25-b3. It is further discussed for hearing in 2.8, 419b18-25, 420a3-19; and for smell in 2.9, especially 421b9-11. Aristotle notes that both these types of perception can take place in water as well as air, but when sounds are heard in water, it still happens that air (in this case, the air inside the ears) plays a role as medium (cf. 420a4-19). The medium of smell, on the other hand, is neither air nor water but rather something 'nameless' (419a32) common to both.

179. See 1059,20-1 above.

180. This passage corresponds to B 245a26-b18.

181. Simplicius is pointing out that this is a case in which there is genuine continuity rather than just touching. See 1054,14-15 above.

182. See 1057,14-16; 1060,18-21 above.

183. This passage corresponds to B 245b19-246a25.

184. Reading *alloioutai* instead of *legetai*, which seems to be merely an editorial slip, since Diels does not note any departure from the text of the *Physics*, and the *Physics* MSS appear to be unanimous in having *alloioutai*.

185. Diels follows the Aldine edition in adding *heteran* here.

186. The present chapter of the *Physics* argues for a much narrower conception of alteration than other passages in Aristotle. For instance, on this more restrictive view the examples of alterations given in *GC* 1.4, 319b12-14 would no longer count as alterations. See Wardy, pp. 163-5; Manuwald, pp. 49-102.

187. Simplicius uses the plurals of these abstract nouns here, indicating that he is referring to particular concrete instances. See the note on the same usage in 1058,1-2 above.

188. 245b6-7 (slightly abbreviated).

189. Aristotle has 'and'.

190. This species of quality is discussed in *Cat.* 8, 9a14-27, which is also the source of the examples. In his commentary on that passage and elsewhere, Simplicius takes these terms to be applied in virtue of the undeveloped ability or disposition rather than the trained skill (see, e.g. *in Cat.* 224,2-3; 242,18-20; 243,8-13.29-31; 245,10-24; 264,16-17), and this is brought out here by the use of the term *epitêdeiotês* ('suitability'). Aristotle, however, says that more than just the mere disposition is needed (*Cat.* 8, 9a16-24), as Simplicius notes in his commentary on the passage (*in Cat.* 247,14-22). Simplicius may be assuming here that the fully developed disposition would count as a *hexis* and hence would be excluded from the scope of alteration.

191. Aristotle observes that the qualities involved in some of these cases do not have names (*Cat.* 8, 10a32-b5). Simplicius' attempt at an explanation notwithstanding, this is one of many puzzles about the relationship of the treatment of qualities here in *Phys.* 7 and their classification in the *Categories*. See 1081,10-30 below.

192. The Greek word *puramis* is usually translated 'candle' in this context, but Wardy (p. 197 n. 52) argues, on the basis of *GC* 2.7, 334a32-4, that it means 'pyramid'.

193. These things are said to be spoken of 'paronymously' because the term used to refer to them is derived from another through a change of ending (*Cat.* 1, 1a12-15). Aristotle is fond of mentioning the changes in description noted here, though he uses it to support different points in different passages. See *Metaph.* 7.7, 1033a5-23; 9.7, 1049a18-24.

194. The examples in this last clause seem to be taken from 245b13-14, but our text of the *Physics* has 'wax' in place of 'wood'. Since wax fits the reference to being 'liquid' better, one may perhaps suspect that the received text of the *Physics* has been corrupted. It is noteworthy that Themistius (Schenkl 205,19) has wood too.

195. That is, whether we say that 'the bronze is liquid' or 'the liquid is bronze'.

196. The reference is to the switching of subjects and predicates in 245b15-16.

197. Of course, we can say the latter in English.

198. 245b16. The term 'homonymously' [*homônumôs*] here simply refers to using the same term. For the technical notion of homonymy, which involves the same term with a different defintion, see 1066,29 and 1085,15ff below.

199. The full version of Aristotle's argument in *Physics* 246a1-4 is: 'Consequently, if with respect to shape and figure, what has come to be is not spoken of as that in which the shape exists, but, with respect to affections and alterations, it is spoken of as [that], it is evident that comings to be cannot be alterations.'

200. *Phys.* 5.1, 225a26-9.

201. Since the protasis seems to lack an apodosis, there may be a lacuna here. The material in brackets supplies the missing conclusion, which Simplicius may view as too obvious to need stating. The Aldine edition solves the problem by changing the *ei* ('if ') in line 22 to *kai* ('and'), but doing so does not seem to offer a complete solution. It is possible that a clause beginning with *dêlon* or *dêlonoti* has been omitted somehow due to the presence of *dêloi* in lines 23 and 25 (cf., e.g. 690,22-3; 850,18-19; 1095,9-11; *in Cael.* 217,32-4).

202. For similar use of the term 'tincture' (*parakhrôsis*), see *in Cael*. 100,20-1.

203. That is, when heated red hot iron will no longer have its natural coldness but will actively heat other things just as fire does. Cf. *in Cael*. 113,15-17.

204. This passage corresponds to B 246a25-9. (Some of the material in A 246a7-8 has in fact has appeared in the previous passage in the B version, specifically in B 245b25-6.)

205. The contrast invoked here, which is that between becoming a different kind of thing and becoming simply the same thing with a different quality, appears in 1081,18-19 below and in *in Cael*. 99,3-4 (cf. 100,20-1). See also Porphyry *Isagoge* Busse 8,17-9,16.

206. 'Being altered' here is *alloioumenon*, which literally means 'being made otherwise'.

207. In this passage 246a10-12 corresponds to B 246a29-30; 246a12-17 to B 246b27-247a20; and 246a17-20 to B 246a25-8 (though this last passage in B actually appears not in the discussion of states, but of shapes).

208. Reading *alla mên oude* in line 26 with the *Physics* in place of the lemma's *alla mên oute*. Simplicius' apparent partial quotation in 1065,5, which has *oude* before *hexeis*, confirms that Simplicius' text of the *Physics* had *oude*. Diels notes in his apparatus that this is what Simplicius read, but he leaves *oute* in the lemma.

209. It is the first species listed in the discussion of quality in *Cat*. 8 (see 8b25-9a13).

210. 'Natural condition' translates *to kata phusin*, literally, 'accordance with nature'.

211. In 1071,16-18 below the corresponding point is made about the virtues of the soul. See also 1081,19-20.

212. The quoted material is taken from 246a13-15, but it has been reordered. Aristotle gives this part of the argument as: 'For, whenever it has acquired its own virtue, then each thing is spoken of as perfect, for then it is most in accordance with nature.' Simplicius rearranges it into the form: 'Whenever it has acquired its own virtue, then a thing is most in accordance with nature. When it is most in accordance with nature, then each thing is spoken of as perfect.'

213. Here and in what follows it is helpful to remember that *teleiotês* also has the sense of 'completeness'. (However, the verb translated 'is completed' in lines 11 and 18 is *sumpe-plêrôtai*, from *sumplêrousthai*.) For other references to this kind of *teleiotês* and use of the corresponding sense of *teleios*, see 501,29-30; 894,18-19; *in Cael*. 8,25-35; 9,13-14; 39,11-49,19 passim; *in Epict*. Dübner 75,53-5.

214. Reading *epeisaktos* in line 14 instead of the apparent misprint *ekeisaktos*.

215. This use of *zôê* in a reified sense is common in later Neoplatonism. In the Simplician corpus, it is especially common in the commentary on the *de Anima* (e.g. 12,25), though the increase in occurrences in that work is amply explained by the subject matter and need not have any bearing on that work's disputed authorship.

216. Homonymy involves use of the same term with a different definition (*Cat*. 1, 1a1-6). In 245b16 (reflected in 1063,10-11 above), the emphasis is on the use of the same term. Here the emphasis is on the difference in the definitions and specifically on the failure of the 'deadened' thing to satisfy the original definition.

217. In this passage, 246b3-14 corresponds to B 246a30-b27, and 246b19-20 to B 247a22-3.

218. *Phys*. 5.2, 225b11-13.

219. Once again these abstract nouns are pluralized as a way of indicating that they refer to concrete instances. See 1058,1-2 above. For more on these 'primary powers', see 1069,20-5 below.

220. Reading just one *kai* instead of the two printed in succession in lines 15-16, presumably by mistake.

221. Parallelism with the *touto* in line 16 suggests that *autê* ('itself') should perhaps be read as *hautê* ('this') here (as in fact the Aldine edition does read it).

222. Aristotle, who seems to have coined the term, uses *homoiomerês* (literally, 'with like parts') to describe things with uniform and homogeneous compositions, so that the parts are like each other and the whole. Flesh, for instance, is distinguished in this respect from a face (see *PA* 2.1). Aristotle also finds special application of the term to Anaxagoras' physical theories (see 1069,24 below).

223. Once again, these abstracts are pluralized, as in lines 12-13 above.

224. *Phys*. 5.2, 225b11-13.

225. In chapters 1 and 2 of *Physics* 5.

226. Diels follows the Aldine edition in reading *to alloiôsis einai* in line 17 instead of *to alloiôsin einai* with MSS AM. In fact, the usual construction in both Aristotle and Simplicius would call for *to alloiôsei einai* with the dative (cf., for example, *to kinêsei einai* in 1043,16). It is possible that the *to alloiôseis einai* found here in MS F can be traced back to that.

227. Reading *allou* in line 22 with the Aldine edition instead of the *alloiou* found in the MSS and printed by Diels.

228. *Phys.* 5.2, 225b11-13.

229. *dokei* has been given its regular translation 'seem' here, but in this context it could have the force 'are generally thought'.

230. Moving Diels's comma from after the phrase *hê eis hugeian metabolê* ('the change to health') to immediately before it. Also, the *gar* in this clause has not been translated.

231. The use of *homoiomerês* to describe Anaxagoras' mixture seems to have originated with Aristotle, who, as noted above, appears to have coined the term (1067,18). According to Anaxagoras, 'in everything there is a portion of everything' (fr. 11 Diels-Kranz; also found in fr. 12), a statement which can be read as implying the requisite homogeneity. Anaxagoras does allow for local predominance of a single kind of stuff, and he also treats *Nous* (Mind) as an exception to the general rule. For Aristotle's use of the term in connection with Anaxagoras, see especially *Cael.* 3.3, 302a29-b5; *GC* 1.1, 314a16-b1. The term and its use are helpfully discussed in W.K.C. Guthrie, *A History of Greek Philosophy*, Cambridge 1962-1981, vol. 2, pp. 281-5, 325-6; and G.S. Kirk, J. Raven, and M. Schofield, *The Presocratic Philosophers*, 2nd ed., Cambridge 1983, pp. 376-8. Simplicius, of course, is the source of many of Anaxagoras' fragments; in connection with the present passage see *in Phys.* 27,9-10; 156,16-17; 164,22-5; 172,1-9.

232. These qualities in fact correspond to the hot, cold, dry, and wet which Aristotle mentions in 246b16-17 (and which have appeared in 1067,12 above). They can be viewed as being more fundamental for him than the elements, each of which is a combination of two of them as form combined with (prime?) matter. See *GC* 2.1, 329a24-36; 2.2, 330a24-9. Important passages in Simplicius on this topic are *in Phys.* 227,26-30 and *in Cael.* 130,18-20.

233. See 1070,12-13 below. The Stoics made all emotional affections bad by nature.

234. The Greek word is *thumoi*. In particular it can be taken as referring to feelings of anger, but in this standard pairing of the so-called 'irrational affections' (*aloga pathê*) it goes beyond just anger. See *in Epict.* Dübner 132,53-4. Plotinus accommodated the Stoic view that the wise person will be free from all affections and the Aristotelian view that we need moderate affections, by treating the Aristotelian ideal as an earlier stage of the spiritual progress, suitable to civic and community life, but not yet to the contemplative life (*Enneads* 1.2 and 1.3.1-4).

235. See 1069,29 above.

236. *Phys.* 5.2, 225b11-13. The other quotations of this text in Simplicius match the version given here (*in Phys.* 409,2-3; 835,13-15; 859,26-7). However, in book 5 Ross, partly on the basis of Alexander's explanation as quoted in 835,1, adds *alêtheuesthai kai mê* before *alêtheuesthai* in 225b13 to give 'to be truly described by a predicate and not be truly described by a predicate'.

237. *Phys.* 5.1, 224b27-28.

238. The first part of this passage, A 246b20-247a7, corresponds in terms of its place in the overall argument to B 246b27-247a23. However, that passage in the B version incorporates material which appears earlier in the A version: compare B 246b27-247a20 with A 246a13-17, and B 247a22-3 with A 246b19-20. (Once again, the reader is referred to Wardy's book for detailed discussion of the differences between the two versions.) In the rest of this passage, A 247a7-13 corresponds to B 247a23-8.

239. Diels, citing 1073,13, prints the second part of the lemma as *autê de ouk estin alloiôsis* ('but itself is not alteration'). (MSS F and M of Simplicius and the Aldine edition have *hautê* rather than *autê*; MS A of Simplicius has *autê* with no breathing mark.) Though *Physics* MSS bcjy match MSS FM of Simplicius and the Aldine edition, Ross, apparently following MSS H and I of the *Physics,* reads *autai d' ouk eisin alloiôseis* ('but themselves are not alterations').

240. cf. 1065,9-24 above (in connection with the bodily virtues) and 1081,19-20 below.

241. 246b19-20. Simplicius has *ê apathes ê pathêtikon hôdi* whereas the *Physics* has *ê apathes ê hôdi pathêtikon*.

242. Reading *ta* in 1071,26 instead of the apparent misprint *to*.

243. In this quotation from 246b17, Simplicius has *prôtôs* ('primarily') instead of Aristotle's *prôtois* ('first'), though Simplicius does have *prôtois* when he quotes this material in 1069,20-2 above (and also in 837,17).

244. *Enn*. 1.1.1.12-13. Simplicius modifies this quotation slightly (including word order). Plotinus says that 'the affections are either certain kinds of perceptions [*aisthêseis tines*] or not without perception'.

245. Both these points are made in 247a1-3. Psychic states are also treated as *teleiôseis* ('perfectings') in 246a10-14.

246. See *EN* 2.3, 1104b8-1105a16.

247. See lines 17-18 above.

248. cf. *EN* 2.3, 1104b22-3; 2.5, 1106b21-2.

249. *Laws* 636D7-E2 (with minor changes).

250. 247a16-17. In incorporating this quotation into his own text, Simplicius omits Aristotle's *de* and adds *eisi*.

251. The subject and verb of this sentence have been pluralized in translation.

252. cf. 1059,1-6 above.

253. This passage corresponds to B 247a28-b21.

254. In MSS AF the lemma has the singular *hexis* whereas Aristotle's text (and Simplicius MS M) have the plural *hexeis*. Simplicius uses the singular form in 1074,6, where he seems to be quoting Aristotle directly, but he uses the plural in his paraphrase in 1073,26-1074,1 (cf. also 1074,20).

255. The text of the concluding part of the lemma in Simplicius is: *epistatai pôs têi katholou ta en merei* ('it somehow knows the particulars through the universal [knowledge]'), and Diels reads exactly the same thing in 1075,5-6. Our MSS of the *Physics* are divided between *epistatai pôs têi katholou to en merei* ('it somehow knows the particular through the universal [knowledge]') and *epistatai pôs ta katholou tôi en merei* ('it somehow knows the universals through the particular'). Ross prints the latter, based in part on what Alexander is quoted as saying in 1075,2-3 below (though in fact Alexander may have had the same text as Simplicius). However, the Neoplatonist Simplicius rejects the empiricism involved in Alexander's approach (see Simplicius' comments in 1075,10-20 below).

256. Reading *oute* in line 26 with MSS AM rather than *oude* with Diels, who follows MS F and the Aldine edition.

257. This reference to habituation (*sunethismos*) seems to be an allusion to the etymological connection Aristotle finds between *ethos* ('habit') and *êthos* ('character'). See *EN* 2.1, 1103a17-18. The commentary on the *de Anima* attributed to Simplicius mentions dogs, apes, bears, and horses as examples of animals which can be trained in this way (*in DA* 308,13-17).

258. Simplicius may be referring to terminology in his own day, or he may be making specific reference to alternative readings in the manuscripts. His wording suggests the latter, but then it is hard to know what passage he might have in mind (unless it is 247b1). In Aristotle, both terms are used for the intellectual faculty of the soul, but of the pair only *dianoêtikê* is used for the psychic state. In Neoplatonism, *dianoêtikê* is used of step-by-step reasoning of the rational soul, *noêtikê* of the higher stage of all at once intuitive understanding on the part of the intellect, which is what step-by-step reasoning aims at. Intellect is higher than reason.

259. 246b11-12. The last two occurrences of 'nor' represent places where Simplicius has *oute* instead of the *oude* found in Aristotle.

260. The phrase *tês hexeôs tês epistêmonos* in lines 12-13, though it can be translated 'of the knowing state' as it has been here, seems a bit unusual. Perhaps we should read *tês hexeôs tês epistêmês* ('of the state of knowledge') with MS F and the Aldine edition or *tês hexeôs tou epistêmonos* ('of the state of the knower').

261. Though the phrase *pros ti pôs* (coupled with the participle *ekhôn*) may be more familiar as the designation of the Stoic category of the 'relatively disposed' (see, e.g. *in Cat.* 66,32-67,2; 67,17-19; 165,32-167,36), Aristotle himself uses the phrase (with the infinitive *ekhein*) to characterize relatives (see especially *Cat.* 7, 8a32, b1). Here in *Phys.* 7.3 Aristotle is simply employing it to refer to the category of relative (cf. *EN* 1.12, 1101b13); the corresponding references to that category in the B version dispense with the *pôs*. See Manuwald, p. 83.

262. Simplicius has 'it' (*auto*).

263. The Greek translated 'an experience [consisting of] cognitions of the individuals' in line 29 is *empeiria tis tôn kath' hekasta gnôseôn*. This might alternatively be taken to mean 'an experience of the individual cognitions'. However, the translation given has the advantage of matching the use of *tôn kath' hekasta* in the next line. (Perhaps a second *tôn* has dropped out right after the *tôn* in line 29.) For the view that *empeiria* consists of several cognitions (or memories), see *An. Post.* 2.19, 100a5-6; cf. also *Metaph.* 1.1, 980b29-981a1, 15-16. For a succinct statement of the view that perception is of individuals but knowledge is of universals, see *DA* 2.5, 417b21-3.

264. The subject and verb of this sentence have been pluralized in translation.

265. This view is in agreement with Alexander's claim in his commentary on the *Prior Analytics* that the 'cognition of the universal is derived from trial of the individuals' (*in An. Pr.* Wallies 332,22-3; cf. also Alex. *DA* Bruns 83,7-8). Earlier in the *in Phys.* Simplicius has quoted Alexander on the role of induction from particulars (53,25-6).

266. As discussed in connection with the lemma, there are two basic versions of the part of 247b6-7 which is quoted here in lines 5-6. One offers an empiricist account of knowledge, giving priority to the role of acquaintance with particulars in coming to know universals; the other emphasizes the role of the universal in knowing the particular. Alexander provides an empiricist interpretation of the second version. Simplicius, notwithstanding Aristotle's support for an empiricist approach in such passages as *An. Post.* 2.19, is about to give a thoroughly Platonic and nonempiricist reading of it. For another view of the passage, see Wardy, pp. 231-4.

267. Reading *epistêmêi* in line 9 for the apparent misprint *epistêmê*.

268. cf. 1079,10-14 below, where Simplicius presents what amounts to a version of the Platonic theory of recollection.

269. *Cat.* 3, 1b15; 5, 2a25-6.

270. 247b6 (in Simplicius' version).

271. This passage corresponds to B 247b21-2.

272. See Aristotle *Sens.* 6, 446b2-4; *Cael.* 1.11, 280b6-9 (along with Simplicius *in Cael.* 313,20-314,14; 315,17-18). Atemporally: without taking time.

273. This passage corresponds to B 247b22-4.

274. This part of *Physics* 7 is making use, although apparently in an early form, of Aristotle's doctrine of the levels of potentiality and actuality. As fully worked out, there are three levels: first, the undeveloped potential; second, the developed state, which is actual relative to the original raw potential but potential relative to the activity corresponding to the state; and third, the full actuality represented by that activity. See *DA* 2.5 (along with *in DA* 119,34-39) and *Phys.* 8.4, 255a30-b24 (along with *in Phys.* 1213,29-1216,20). Some of the terminology regularly associated with this doctrine is present here in *Phys.* 7.3, but the doctrine itself is not explicitly formulated. Put in terms of the later explicit version, the point is that not only does the change to the active exercise not take place through coming to be or alteration, but the same is true of the change to the state. This claim is in line with the discussion in *DA* 2.5, though Simplicius here seems to be talking about a stage involving preparation for acquiring the state. There is no mention of that in the *DA*, but Simplicius may be trying to carve out a niche for the 'settling down' which figures so prominently here.

275. This is not formally proved, but cf. *Phys.* 5.6, 230a4-5.

276. *Phys.* 5.2, 225b15-16 (first clause slightly modified). R. Sorabji notes in personal communication that this passage implies Aristotle's rejection of acceleration.

277. Simplicius is quoting from 247b12-13, but the first part of his quotation differs from the *Physics* text Ross prints. In place of Simplicius' *geneseôs gar oudemia metabolê* ('For there is no change of coming to be'), Ross has: *holôs gar oudemias metabolês* ('For [there is coming to be] of no change at all').

278. Simplicius is referring to what Aristotle says in 247b11-12: 'for it is because the mind is at rest and has come to a halt that we are said to know and to understand.' The etymological claim is that *epistasthai* ('to know') is derived from *histasthai* ('to come to a halt').

279. *Tim.* 43A4-7.

280. *Tim.* 44A7-B6 (with minor changes).

281. Simplicius has the adjective *katamelês* in line 22 whereas Plato has the aorist active participle *katamelêsas* (44C2). The adjective, which is not found in Liddell-Scott-Jones, *Greek-English Lexicon* (or its 1968 supplement), appears here in the Aldine edition; Diels notes that he is relying on the silence of his collator as to its actual occurrence in the MSS.

282. *Tim.* 44B8-C4 (with minor changes).

283. *Phys.* 5.2, 225b33-226a6.

284. The text Simplicius is discussing here is a version of 247b10 reading: 'The original acquisition of knowledge on the one hand [*men*] is not coming to be.' The unanswered *men* is found in MSS HI of the *Physics* but is omitted by Ross.

285. That is, this alternative text which Simplicius mentions has *de* ('on the other hand') in place of *gar* ('for') in 247b11.

286. This passage corresponds to B 247b24-248a27.

287. The lemma omits the *hotan* ('whenever') which Aristotle's text has after *hôsper* ('just as').

288. *Enn.* 1.1.9.15. Cf. 1075,15 above. This use of *prokheiron* goes back to the *Theaetetus* (198D7, 200C2).

289. Simplicius provides more details about this largely Platonic view in 1249,32-1250,4 (a passage which shows how non-Aristotelian it is). Cf. also *in DA* 121,11-126,16 (where there are references to *Phys.* 7); 308,33-7.

290. The verb is *prokheirizesthai*, which is cognate with *prokheiron* ('ready to hand').

291. The *ho prokekheirismenos* ('person who has made it ready') in line 22, which Diels takes from MS M, is translatable but awkward, though it is parallel to *ho energôn* ('the one who is active') in line 25. The noun *prokekheirismos* ('advance readiness') found in MS A (and the Aldine edition edition) is attractive, but the word is seemingly unexampled elsewhere. (It may, however, be supported by the *kekhôrismos* in MS F.) Fortunately, the general sense is not in doubt.

292. This passage corresponds to B 248a27-b27.

293. 248a4. Simplicius has *tinôn en tôi sômati* ('certain things in the body') here in line 14 (and also in line 21 below), whereas the *Physics* has *tinôn tôn en tôi sômati* ('certain of the things in the body'). Simplicius does have *tôn* after *tinôn* in the quotation of the same material in 1078,4-5. (The *tôn* is also missing in 1081,4, where Diels adds it.)

294. Omitting the unnecessary *tôn* which Diels and the Aldine edition insert before *sômatôn* in line 16.

295. Reading *lambanein* in line 22 instead of *analambanein*, which in this context would have to mean 'reacquire'.

296. This passage corresponds to B 248b27-8.

297. cf. *DA* 1.4, 408b2-18.

298. The *tôn* is added by Diels.

299. Simplicius, as often, uses the singular *kath' hauto* ('in its own right') even though it is meant to apply to the plural 'relatives'.

300. *Cat.* 8, 8b26-9a13.

301. This is a quotation from *Cat.* 8, 10a11-12. Simplicius (or Alexander) omits *huparkhousa* before *morphê.*

302. See 1064,15 above.

303. The *einai* ('to be') could simply be read with *hoper legetai* ('the very thing it is said'), but the word order is against doing so here (though it encourages it in 1071,17). See 1065,9-24 and 1071,16-18 above.

304. The material in brackets is based on the fuller discussion in 1058,6-8 above.

305. Some of the wording here is taken from 243a32-4.

306. cf. *in Cael.* 422,34.

307. Except for *sunexetazein*, which does not appear in the Aristotelian corpus, Aristotle uses all these verbs with the meaning 'compare', though in this chapter of the *Physics* he uses only the verb *sumballein* (cognate with *sumblêtos*) and that only once (249b5). The translations provided in the present passage are intended to distinguish the different verbs as well as to reflect their respective etymologies. However, when Simplicius uses these words later in his commentary to talk about comparison (and he uses of all them except *sunexetazein*), they will often simply be translated 'compare'. The Greek-English Index provides citations and translations for all occurrences.

308. The idiomatic but pleonastic *eis tauto* ('into the same thing' or 'into the same place') has been omitted in translation.

309. Aristotle uses this claim in *Phys*. 6.2, 232b14-17, but he does not in fact prove it.

310. cf. 1096,7-9 below, where Alexander suggests that in Aristotle's time the problem was still being treated as an open question. (The verb *apegnôsto*, from *apoginôskein*, here translated 'had been given up on', could perhaps be taken to mean that the claim of equality 'had been rejected'.)

311. Simplicius discusses attempts to square the circle, i.e. to construct with ruler and compasses a square equal in area to a given circle, in a long section of *in Phys*. running from 54,12 to 69,34, and he also touches on it in *in Cael*. 412,30-413,2 and *in Cat*. 192,12-30 (commenting on *Cat*. 7, 7b31-3).

312. See 249b3. Elsewhere, see *Cat*. 6, 6a26-35; 8, 11a15-19; *Metaph*. 5.15, 1021a11-12. cf. the Subject Index s.v. 'equality' and 'similarity'.

313. The sense seems clear enough, but with *oud' an* one would expect an optative, either *eiê* or perhaps *dunaito* in place of the *dunaton* in line 14 (cf. 75,5-6; 112,4-5; 1246,24-5; *in Cael*. 235,33-4). However, the omission of the verb is not unexampled (see 152,5).

314. Reading *tou touto einai* in line 19 instead of *toutou to einai*.

315. 248a16. Simplicius inserts 'time' (*khronôi*).

316. Simplicius seems to be referring to 248a17-18, where, however, in place of *ê elattôn* ('or smaller'), our text of the *Physics* reads *oud' elattôn* ('nor smaller').

317. The sequence *pote men ... pote de* is so regular that it seems better to postulate homoeoarcton and read *pote men meizôn* in line 16 with MS F (even if that is only a conjecture in F) than to follow Diels in reading *pote meizôn* with the other MSS.

318. See 1038,2-3 above.

319. Simplicius is quoting from 248a25-b1 but also inserting additional words for clarification.

320. See *Phys*. 6.2, 232a25-6; cf. ibid. 4.14, 222b31-223a4.

321. cf. *Phys*. 6.2, 232a26, b5-6

322. 248b3-4. Simplicius has *estai ti meros tou A en hôi to B tou kuklou dieisi*, which is found in *Physics* MSS EHIJK. Simplicius takes *tou kuklou* with *to B* to give 'B belonging to the circle'. Ross, on the other hand, reads *to ison* before *dieisi* in 248b4 with *Physics* MSS Fbcjy. Given that reading, it would be more natural to take *tou kuklou* with *to ison* and to take *to B* as referring to the thing moving over the circle. In Ross's version, then, the entire sentence in the *Physics* reads: 'There will be some part of the [time] A in which B will go through [an amount] belonging to the circle equal to [what] C [will go through moving over the straight line] in [time] A as a whole.'

323. 248b4. Though Simplicius reads it, Ross brackets 'the [line] C' (*tên G*) as a gloss. Aristotle has complicated matters by apparently using the letter 'C' both for the moving object and the line. In the Greek, however, the gender of the definite articles distinguishes which is meant.

324. Simplicius remarks in 1086,20-5 below that this sentence was transmitted in different ways: as *all' hosa mê homônuma hapanta sumblêta* ('but absolutely all those things which are not homonymous are comparable') in some places and as *all' hosa mê sunônuma hapanta asumblêta* ('but absolutely all those things which are not synonymous are noncomparable') in others. Ross prints *all' hosa mê sunônuma, pant' asumblêta* ('but all those things which are not synonymous are noncomparable') and traces the corruption to a misreading of *pant' asumblêta* ('all are noncomparable') as *panta sumblêta* ('all are comparable'). According to Ross, that error in word division led in turn to a compensating change of *sunônuma* to *homônuma*. See his note on 248b6-7 (p. 678).

325. This sentence follows the version of the text given in the lemma, which is different from Ross's text.

326. Things sharing the same term but with different definitions are said to be 'homonymous'; things sharing the same term with the same definition are 'synonymous'. For the distinction, see *Cat.* 1, 1a1-12 (and also what Simplicius says in 1096,26-7 below). Aristotle has used *homônumôs* earlier in 245b16 when noting that in cases of alteration we designate the matter with the term for the affection (see 1063,10-12 above). The emphasis there, however, seems to be simply on the use of the same term and not, as usual, on the difference in definition.

327. 248b8. Simplicius has turned Aristotle's wine (*ho oinos*) into vinegar (*to oxos*)!

328. Literally, 'sameness of fastness' (*to tauton tou takheos*).

329. 249a23-4. See 1096,28-1097,6 below.

330. The alternative readings are discussed in the footnote to the lemma.

331. This text, which would be from the 'B version' of chapter 4, is not found in the surviving MSS of the *Physics*. What Simplicius says in 1036,4-6 implies that there were two versions of the whole book, but we only have a second version for the first three chapters. For more details see n. 13 above.

332. Ross (p. 678 *ad* 248b13-14) and Wardy (p. 281 n. 19) take the meaning of *dunamis* as used by Simplicius in this passage to be 'weight'. Support for this view comes from Simplicius' connecting *dunamis* with the essence and form of air and water in 1087,7-8.24-5 and 1089,27-9, since Aristotle does tie the essence of the elements to tendencies to move up or down (that is, roughly speaking, to weight as he understands it). However, the reference here in Simplicius' rather speculative interpretation of Aristotle's meaning in this passage might be to something else, e.g. cooling capacity (see *GC* 2.6, 333a23-7; cf. *in Cael.* 81,24-82,8). Accordingly, *dunamis* has simply been translated in a neutral way as 'power', with the proviso that it may mean 'weight' (cf. *in Cael.* 676,23-7; 677,5-24).

333. Substituting a period for Diels's comma here.

334. The meaning seems clear, but Diels's making the *ho* in line 32 the relative pronoun rather than the definite article seems unlikely to be the whole solution to whatever may be ailing this passage. An anonymous reviewer, citing 1088,28-9, offers the conjecture *allôs ge* for *all' oukh ho*. Perhaps the original wording was *all' allôs*, and the *allôs* dropped out, leading to further changes.

335. A long sentence (which includes phrases quoted from 248b14-15) has been broken up.

336. Literally, 'is said of both the things which the term too [is said of]'.

337. cf. 1098,27-8 below. For definitions of equality in terms of being measured by the same measure (or measures), see 614,25 and *in Cat.* 151,18-19; 153,21-2.

338. This would read better if it were 'the primary things which have received them'. Bodies (*sômata*) may even be an intrusive gloss; in any case, the word is more relevant to magnitude, which is considered later (in 1089,21ff). 'Primary' (*prôtos*) is used here to pick out the proximate recipient or bearer of a property; for instance, in a white horse the whiteness belongs 'primarily' to the surface of the horse and only secondarily to the horse (cf. 1090,25-30 below).

339. These same two examples of white things appear in *EN* 1.6, 1096b22-3.

340. The Greek word *leukos*, in addition to meaning 'white' (or 'pale'), can also mean 'clear'. See 1098,7-9 below.

341. The adverb *entreptikôs* ordinarily means 'reproachfully' or 'reprovingly', but on the two other occasions when Simplicius uses the word it seems to combine that meaning with 'decisively' (cf. 329,20; *in Epict.* 137,8). Thus for him it may mean 'so thoroughly as to put to shame', and 'damningly' is used here as a way of conveying that double meaning.

342. Literally, 'to eliminate the nature of homonymous things', where 'the nature of X' is a periphrasis for 'X' (see LSJ, s.v. *phusis*, II.5). The same usage in Aristotle is noted in Bonitz, *Index Aristotelicus*, Berlin 1870, s.v. *phusis* (at 838a8-12). An anonymous reviewer suggests that *phusis* may mean 'class' here (cf. LSJ, s.v. *phusis*, VI).

343. 249a2-3. Simplicius has *ouk esti to tukhon tou tukhontos dektikon* in lines 16-17, whereas the received text of the *Physics* is *dektikon ou to tukhon estin*. On the strength of

Simplicius' wording here and in 1096,2 Ross inserts *tou tukhontos* after *tukhon* in his text (see his note on 249a3, p. 680).

344. This phrase translates *to prôton* in 249a3. The meaning is that the primary recipient in particular is not capable of receiving just any property. See lines 24-30 below.

345. Instead of 'taste' (*geusis*) one would expect Simplicius to mention some kind of substrate here (perhaps *to geuston*). It is possible that *geusis* refers to the organ of taste (which is a substrate of a sort), just as, depending on the context, *aisthêsis* can refer both to sense perception and to the sense organ. Though the word *hugron* ('liquid') found in MS M offers an attractive alternative, it seems likely to be a conjecture intended to mend what appears from Diels's apparatus to be a somewhat troubled spot.

346. For *khrôma* as a musical term, cf. *in Cat.* 187,5-6.

347. Punctuating with a full stop.

348. Closing the first parenthesis after *leuka* in line 20. Diels has only a single long parenthesis which he closes after *sunkrinomenai* in line 21, but his punctuation destroys the parallelism of the *mête ... mête* construction in lines 20-21. There are in fact two parenthetical elements in this sentence, each introduced by *hoion* ('for instance'), and Diels's punctuation has been changed accordingly.

349. cf. *in Cael.* 390,5-8. Alexander, on the other hand, does use 'length' [*mêkos*] in a case which does not involve a straight line (quoted in 1308,20-1).

350. In this and the following sentence, Simplicius is quoting and paraphrasing 249a13-15. He refers to Aristotle's question again in 1095,33-4.

351. 249a15-16. The MSS of Simplicius have *kat' ekeina eidei diapherei* ('differ in species in accordance with those'). The MSS of the *Physics* offer several different readings; Ross prints *kai ekeina eidei diapherei* ('those too differ in species'). Diels indicates that Simplicius wanted that same text here, pointing to Simplicius' quoting those words in 1093,7. However, Simplicius attributes that version to 'different manuscripts'.

352. Reading *anthupopherei* in line 17 instead of *hupopherei*. The latter word does not seem to fit here, but Alexander uses the former in the sense of 'reply' in *in Sens.* Wendland 87,3. (It is possible that Alexander has used *anthupopherei* just before the quoted extract and the shorter form *hupopherei* simply serves to repeat it.)

353. Reading *isos* in line 17 instead of *isôs*; the latter may be due to *isôs* in line 18.

354. Simplicius' views about how this statement is to be interpreted also come out in 1092,6-12. Ross, in his note on 249a9-10 (p. 680), agrees with Simplicius' taking *en tôidi* (literally, 'in this') to refer to the length, though he admits that the construction is difficult. Wardy (p. 286 n. 24), on the other hand, argues that *en tôidi* should be taken as referring to time, and that does seem more plausible.

355. Reading *hoté* (indefinite) with Ross in place of the lemma's *hoti* ('that'). It is hard to tell what Simplicius read, but the text of the lemma as printed is very awkward.

356. Simplicius is glossing *to auto megethos* (Ross: *tauto megethos*) in 249a19. However, in the combination of quotation and paraphrase which he has given in the previous sentence, Simplicius has substituted *diastêma* ('distance') for Aristotle's *megethos* ('magnitude').

357. Reading *zêtômen* in line 28 instead of *zêtôi men*.

358. 249a2-3. For the first clause, Simplicius has *ou to tukhon tou tukhontos dektikon*, whereas Aristotle (with Ross's supplement) has *dektikon ou to tukhon <tou tukhontos> estin*. See the note on 1090,16 above.

359. cf. 1082,25-1083,3 above.

360. The point seems to be that Aristotle's analysis is a fully general one and does not hinge on whether straight lines and curves can be equal.

361. Aristotle uses the phrase *mia phusis* in 248b21. Here Simplicius could mean 'because there is a single nature', but in that case one would expect the accusative instead of the nominative.

362. One could read *en tôi eidei* ('in species') in line 16 with Diels (following MSS AM and the Aldine edition), but reading *hen tôi eidei* ('one in species') with MS F seems preferable in view of the context and involves merely a difference in breathing. (The definite article with *eidos* is no obstacle to this reading, as is shown by the occurrence of *hen tôi eidei* in 113,30 and 114,8.)

363. 249a22-3. Simplicius has *polla lanthanei*, Aristotle the reverse. '*Para touto*' in 249a22 is difficult. It could mean 'besides this (sc. case)', 'present in this (sc. genus)', or 'because of this (sc. the failure of the genus to be unitary)'. Simplicius seems to take it in the last sense, and it has been so rendered here.

364. The text of Simplicius has *homônumôn* in line 28, though it has *homônumiôn* ('of homonymous meanings') in the quotation of Aristotle's sentence in 1086,19 and that should possibly be read here (with the Aldine edition). (However, *homônumôn* appears in the paraphrase in 1114,24.)

365. 249a23-4. Simplicius uses the finite verb *apekhousin* instead of Aristotle's participle *apekhousai*.

366. 249a24. Simplicius uses the finite verb *ekhousi* instead of Aristotle's participle *ekhousai*.

367. 249a24-5 (Simplicius adds *eisin*). For Aristotelian use of these distinctions, see *EN* 1.6, 1096b25-9. In Simplicius, see *in Cat.* 31,22-33,21; cf. *in Phys.* 94,28-9; 346,12-13; 1114,23-5. See also Owen, 'Logic and metaphysics', pp. 179-81.

368. cf. *in Cat.* 74,30-1; 221,2-7.

369. cf. *in Cat.* 31,32-32,3.

370. Simplicius is referring to 249a9-12.

371. Simplicius takes these words from 249a11-12, though he has modified the construction to suit his context. 'Because' translates *aition*.

372. Simplicius will go on here to qualify the claim that motion is a genus. Elsewhere he notes that it is not one (e.g. 402,20-405,23 passim; 449,26; 882,10-13).

373. The *Physics* has *ei* ('if') in line 20 instead of the *hêi* ('insofar as') which Diels reads largely on the basis of a correction in Simplicius MS A. Diels also reads *hêi* in 1101,1 though the Simplicius MSS are divided. Here, since the *Physics* MSS seem to be unanimous and since what Simplicius says in 1100,34-1101,5 suggests that he too may well have read *ei*, it seems justifiable to emend the lemma.

374. Ross, following Pacius, adds *ison ê* in 249b19 to give 'but equal or unequal if that is <equal or> unequal'.

375. Aristotle has 'When, then, ...?' (*pote oun*), as correctly given in the lemma.

376. 249a2-3. See 1090,16-18; 1096,2-3 above.

377. As noted above in connection with 1090,5-9, the Greek word *leukos* means both 'white' and 'clear'.

378. Simplicius is probably thinking primarily of Protagoras as portrayed in the *Theaetetus*. Though Aristotle does not mention Protagoras in *Physics* 7, Simplicius refers to him again, in connection with the millet seed argument in 1108,19-28, where he is once more presented as a champion of sense perception. There are no other explicit references to Protagoras in Simplicius' commentary on the *Physics*, but he is mentioned in *in Cael.* 293,2 and *in Epict.* 131,14 (though not in connection with any particular epistemological views).

379. Diels punctuates this sentence as though it were a question (he puts the question mark right before *epeidê*, 'since', in line 16). On this interpretation, in lines 11-17 Simplicius would simply be posing, in the form of two questions, the alternatives of following sense perception or following reason. However, when *ê* (literally, 'or') follows a question, as it does in line 14, it standardly introduces the commentator's reply. Hence, it seems better to replace Diels's question mark in line 16 with a comma. Diels may have based his punctuation on Simplicius' statement in line 18 that Aristotle has posed puzzles, but though Aristotle in effect asks whether we should follow appearances he does not set up any sort of dilemma.

380. 249a30-1. Simplicius has *ton men takheôs iathênai ton de bradeôs* whereas Aristotle has *ton men takhu ton de bradeôs iathênai*.

381. Of course, Aristotle argues against this view in chapter 3 (see especially 246b3-20). This may be evidence that at least chapters 3 and 4 of book 7 were not written together.

382. The verb *ên* is actually the so-called philosophical imperfect and refers back to the previous discussion.

383. cf. 248a11-12.

384. cf. 1088,23-4 above.

385. *Cat.* 8, 11a15-19; cf. 6, 6a30-5. See the Subject Index s.v. 'equality' and 'similarity'.

386. Simplicius is quoting and paraphrasing 249b4.

387. Literally, 'to the healths which are equal in speed'.

388. 249b6-7. Simplicius changes Aristotle's finite verb *huparkhei* to the infinitive *huparkhein* and accordingly has *mête mallon mête hêtton all' homoiôs huparkhein* in place of Aristotle's *oute mallon oute hêtton all' homoiôs huparkhei*. *Homoiôs* ('to a similar degree') here has the force of 'equally' or 'to an equal degree'.

389. The genitive plural *poiôn* in line 18, which is translated 'qualities', could also be taken to mean 'qualified things', and correspondingly *posôn* in line 17 could be taken to mean 'things having some quantity'.

390. Without the aspiration, the word would be the present subjunctive 3rd person singular form of the verb *einai* ('to be'). With the aspiration, the full clause in 249b9-10 reads: 'inasmuch as these already produce [*poiei*] species of alteration.' Simplicius has just quoted the verb form *poiei* in line 20, but there he gives it a singular subject. That change in subject from singular to plural did not require a change in the verb, since in Greek neuter plural subjects often take singular verbs. (The situation is different in English, however, and hence in line 20 the translation has 'produces' instead of 'produce'.)

391. See *Metaph*. 5.9, 1018a12-13; 10.3, 1054b25-1055a2. These ways in which things can be different correspond, of course, to the ways in which they can be the same or can be one. Aristotle has already addressed the unity and individuation of motion in 242a66-b42 (for Simplicius' comments on that passage see 1043,9-1044,16 above).

392. The *genei* Diels takes from MSS AF has been retained here, but one could also read *genê* with MS M and the Aldine edition. The last clause would then be: 'which indeed are different genera of being.'

393. In this and the following sentence Simplicius is quoting and paraphrasing 249b12-14.

394. Based on what he says in lines 13-18, Simplicius seems to take the phrase *kath' hauto kai mê kata sumbebêkos* with *eidei diapherein dein* to give: 'must differ in species in their own right and not incidentally.' On the other hand, it is arguable that Aristotle intends the phrase to go with *hôn eisin hai kinêseis*, with the meaning: '[the things] to which the motions belong in their own right and not incidentally.' However, the discrepancy, if there is one, would not change the point of this truly tangled sentence.

395. Correcting *sumbekêkos* in line 17 to *sumbebêkos*.

396. See the Subject Index s.v. 'sameness' and 'similarity'.

397. Reading *ei ison* ('if [it is] equal') in line 1 with MS M and the Aldine edition instead of Diels's *hêi ison* ('insofar as [it is] equal'). The former may well be, as Diels suggests, a later correction based on the *Physics*, but there is no unanimity in the Simplicius MSS (MS A has *ê ison*, MS F *kai ison*). More importantly, what Simplicius says seems to suggest that he read *ei* rather than *hêi*. See 1097,20 above.

398. *t'auta* ('the same') in line 2 is printed without the smooth breathing mark which should be above the upsilon.

399. The quoted words are from 249b17-19.

400. This *ti* is added by Diels.

401. 249b24. In 249b23-4 Aristotle seems to be making somewhat *ad hoc* use of the Pythagorean identification of things with numbers. It appears to have been part of this approach that numbers were grouped into species, perhaps on the basis of the shape of their geometric representations. For instance, 9 and 16, which are both 'square' numbers, would be of the same species and hence could be compared. On this approach, a substance corresponding to 16 which came to be in the same time as a substance corresponding to 9 did would have come to be 'faster'. We are not given much information about the different species of numbers, though in 1102,9 Simplicius does describe some numbers as 'human'. See Ross's note on 249b22-6 (pp. 682-3). (Ross revised this note after its initial publication to endorse Jaeger's view that Aristotle's appeal to such number doctrine supports an early dating of *Physics* 7. For discussion of this passage in the *Physics*, see W. Jaeger, *Aristotle: Fundamentals of the History of his Development*, R. Robinson trans., 2nd ed., Oxford 1962, p. 297 n. 2. However, Wardy [p. 290], who draws attention to Ross's revision of the note, is right to be cautious about using such apparently offhand examples as evidence for Aristotle's own views.

Simplicius himself is aware that Aristotle can use an example which does not really reflect his own views [*in Cael.* 454,13].)

402. This passage shows that 'different' might be a better translation for *heteros* than 'of another kind', but in this passage 'different' has generally been reserved for *diaphoros* (see the Greek-English Index).

403. In *in Cael.* 9,10-11 Simplicius notes that it is unusual for Aristotle to make use of Pythagorean 'proofs' (*endeixeis*).

404. From the so-called *Hymn to Number*. Simplicius also quotes this line in 453,12. It is also quoted in Asclepius *in Metaph.* (*CAG* 6.2) Hayduck 38,19.

405. Simplicius also quotes this line in *in Cael.* 580,16. Apparently quite well known, it is found elsewhere in Heron Alexandrinus *Def.* Heiberg 138,9,3 (from Anatolius); Iamblichus *de Vit. Pythag.* 162 (= Dübner 91,13); Philoponus *in Phys.* (= *CAG* 16) Vitelli 388,1; Plutarch *de An. Procr.* 1029F; Sextus Empiricus *Math.* 4,2; 7,94 & 109; Syrianus *in Metaph.* (= *CAG* 6,1) Kroll 103,21-2; 122,33-4; Themistius *in Phys.* (= *CAG* 5,2) Schenkl 79,26; idem, *in DA* (=*CAG* 5,3) Heinze 11,27; Theon Smyrnaeus *de Util. Math.* Hiller 99,16. See H. Thesleff, *The Pythagorean Texts of the Hellenistic Period*, Åbo 1965, p. 173; and W. Burkert, *Lore and Science in Ancient Pythagoreanism*, trans. E.L. Minar, Jr., Cambridge Mass. 1972, p. 73.

406. Literally, 'be constituted', *sustênai*.

407. The second part of the lemma agrees with some MSS of the *Physics* in being fuller than the version printed by Ross.

408. Reading *houtos* in line 29 with MS M instead of *autos*. The phrase *houtos ho logos* ('this discussion') seems to make better sense here than *autos ho logos* ('the discussion itself'). It is also a much more common phrase: a computer search shows that, including quotations, it occurs elsewhere in Simplicius 25 times, as opposed to a single occurrence of *autos ho logos* (*in Cat.* 256,7). The way in which *houtos ho logos* both fits the context and also conforms to Simplicius' regular usage seems to overcome any advantage *autos ho logos* might gain from being the *lectio difficilior*.

409. The relevance of these results extends further, since Aristotle also uses them (though without making specific reference to the discussion here) in his proof in book 8 that the unmoved mover must be without magnitude and cannot have only finite power. See in particular *Phys.* 8.10, 266a12-b24 and, in Simplicius' commentary, 1321,3-1326,37.

410. This is case 4, which is presented in 250a6-7.

411. See 250a12-19.

412. Aristotle makes this point again in *Phys.* 8.5, 256a22. See 1226,25-6 and the quote from Alexander in 1222,27-8.

413. sc. that a mover imparts motion 'up to some length'.

414. This is proved in *Phys.* 6.6, 236b32-237a17.

415. Motion, like all continuous magnitudes, is infinitely divisible. For motion as continuous (in the sense relevant here), see *Phys.* 4.11, 219a11-14; 4.12, 220b24-8. For the infinite divisibility of what is continuous, see *Phys.* 1.2, 185b10-11; 3.7, 207b16-17; cf. 6.2, 233a24-8; 6.4, 235a13-b5.

416. This is proved in *Phys.* 6.5, 236a13-27.

417. The phrase 'of some quantity' [*poson ti*], which Simplicius uses in line 29 and again in line 30, is used by Aristotle in both 249b29 and b30.

418. The Greek word translated 'some particular amount' is *tosonde*, which generally means 'this much' (as, e.g. in line 10 above) or 'this amount'. Here, however, it appears to be used in an indefinite (albeit particularizing) sense. (In *Phys.* 5.1, 224b6 Aristotle uses the phrase *to tosonde* as a general term for quantity.)

419. Thanks to an anonymous reviewer for pointing out that *diastêmati* in line 32 should be *diastêma ti* (as above in line 14).

420. The article with 'D' is feminine rather than masculine or neuter as one might expect (the Greek word *khronos* is masculine and Aristotle often prefixes the neuter definite article to letters). Such feminine forms are a feature of these examples in both Aristotle and Simplicius, apparently because the quantities are viewed as being represented by lines and *grammê*, the Greek word for line, is feminine. See Simplicius' comments in 936,5-6 and 1005,20-1, and see in addition Ross's note on 250a12-13 (pp. 685-6).

421. The talent as a unit of weight varied, but it can be taken to be about 57 pounds (26 kilograms). A stade was about 607 feet (185 metres).

422. Simplicius is using *enallax* ('alternately') in its technical mathematical sense. A passage from the discussion of justice in the *Nicomachean Ethics* gives a very clear illustration of this meaning: 'It will be the case, therefore, that as the term *a* is to *b*, so *c* is to *d*, and, alternately [*enallax*], therefore, as *a* is to *c*, *b* is to *d*' (*EN* 5.3, 1131b5-7). There are additional references to alternate proportions in *An. Post.* 1.5, 74a17-25; 2.17, 99a8-11. Aristotle would have been familiar with such proportions from the work of Eudoxus. See T.L. Heath, *Mathematics in Aristotle*, Oxford 1949 (rpr. 1970), pp. 43-4. Eudoxus' work is generally taken to be the basis of the treatment of proportion in book 5 of Euclid's *Elements* (on alternate proportions see book 5, definition 12 and proposition 16). In this instance, Simplicius gives the ratios in the reverse of the usual order; that is, instead of a:c = b:d, he gives them as b:d = a:c. See also *in Cael.* 233,26-7; 235,9-11; 593,13-16; *in DA* 153,10-11 (a nonmathematical application); 272,18-20. Simplicius is reported in the Arabic tradition to have been a mathematician and to have written mathematical works; see I. Hadot, 'La vie et l'oeuvre de Simplicius d'après des sources grecques et arabes' in id., ed., *Simplicius, sa vie, son oeuvre, sa survie*, pp. 20, 36-8 (or in translation, 'The life and work of Simplicius in Greek and Arabic sources' in Sorabji, ed., *Aristotle Transformed*, pp. 288, 301-2).

423. The apodosis in this conditional matches case 1 (the protasis being a repetition of case 2), so in effect what Alexander has done is to reverse the order of the first two cases.

424. It appears that the last clause should really be: 'so is an hour to two stades.'

425. In the absence of the Alexander passage it is difficult to be certain, but Alexander's presenting case 2 first and then case 1 seems to have led to some confusion. Apparently so has Alexander's treating case 1 as in some sense following from case 2. When Simplicius says he does not find Alexander's case in Aristotle, he may be referring to the alleged implication between case 2 and case 1, since there Alexander is keeping the force and the weight (i.e. half the original weight) constant but doubling both the distance and the time. Aristotle does not address that case directly in connection with locomotion (though in 250b2 he does so in connection with alteration).

426. In this instance, Simplicius turns a:b = c:d into c:a = d:b. The latter is not the standard form of the alternative proportion, which would be a:c = b:d.

427. The phrase *tês A* ('of A'), which is also found in *Physics* MSS bcjy, is not printed by Ross.

428. The summary reflects Simplicius' text of the *Physics*. Modern editors follow most *Physics* MSS in reading *ouk anankê* ('it is not necessary') in 250a10 rather than *anankaion*. In 1106,1-6 Simplicius quotes the version with *ouk anankê* and acknowledges its superiority, since with it Aristotle no longer seems to be denying in 250a13-16 what he has just stated in 250a9-12 (cf. what Simplicius says in 1105,25-31 and 1106,11-12). Ross has a full discussion in his note on 250a9-15 (pp. 684-5). Cornford offers a way of understanding the passage without the negation. See the explanatory note in the Loeb edition (P.H. Wicksteed and F.M. Cornford, trans., *The Physics*, London-Cambridge Mass. 1934, vol. 2, pp. 258-9), and also F.M. Cornford, 'Aristotle, *Physics* 250A9-19 and 266A12-24', *Classical Quarterly* 26, 1932, 52-4.

429. The phrase *tou mêkous* ('the length') appears here in the lemma but is not in any *Physics* MS. It is also not in the Aldine edition and is missing from Simplicius' quotation of an alternative version of the same sentence in 1106,1-4.

430. From 250a8-9, but Simplicius has added the letters 'A' and 'B'.

431. Simplicius' account of case 5 reflects his text of 250a9-12.

432. This text of 250a9-12 (with minor variations) is found in the majority of the MSS of the *Physics*. As indicated already, Simplicius' text read 'it is necessary' instead of 'it is not necessary'.

433. For ease of reference, the five cases as Simplicius has presented them so far are: If force A moves weight B the distance C in time D,

(1) A moves B/2 the distance 2C in time D (250a1-3);
(2) A moves B/2 the distance C in time D/2 (250a3);
(3) A moves B the distance C/2 in time D/2 (250a4-5);
(4) A/2 moves B/2 the distance C in time D (250a6-7); and

(5) A/2 moves B the distance C/2 in time D (250a9-12 as read by Simplicius).

434. Diels writes *ei de* ('But if ') here in the lemma in agreement with Simplicius MS F and *Physics* MSS EK. However, the reading which is better attested for the *Physics*, and the one which Ross adopts, is *ei dê* ('If, then, ...'). There is no positive indication in Diels's apparatus what reading the Simplicius MSS other than F have.

435. In the *Physics*, Ross reads the present *kinei* in 250a12 instead of the future *kinêsei* which appears in the lemma. Both readings are found in the *Physics* MSS, but Ross (p. 685) argues that in this passage Aristotle seems to be deliberately using the present or aorist in the protasis of the conditionals and the future in the apodosis.

436. For this case, involving half the original force and exactly the original weight, see 1105,15-17 above.

437. Simplicius has been led by his faulty version of 250a9-12 to adopt an implausible explanation of what Aristotle meant.

438. Simplicius is following his text of 250a15, but in strict logic the order of 'A' and 'F' should be reversed; that is, the last part of the sentence should read: 'some portion of the distance C which will be in the same ratio to C as a whole that the portion of A, which he named F, is to A.'

439. Previously, in 250a7-8, 'F' (used in the translation for the Greek letter zeta) was assigned to half the weight B. Simplicius seems to have had some additional material in his version of 250a12-19 in which Zeta was assigned to an indefinite part of the moving force. (Traces of this material may survive in *Physics* MSS EJK, which have zeta instead of Epsilon in 250a15.) In 1108,3-4 Simplicius criticizes Alexander's proposal to retain the original assignment of 'F' to half the weight.

440. The 'so that' (*hôste*) is added by Diels.

441. Alexander seems to be quoting (or paraphrasing?) a version of 250a14-15 which has 'so F is to A' instead of the 'as A is to E' printed by modern editors (including Ross). It is difficult to say just how extensive the differences between Alexander's text and the modern one are. The discrepancies may be related to the substitution of *anankaion* for *ouk anankê* in 250a10, but there also seems to be some confusion involving the use of the Greek letters epsilon and zeta ('E' and 'F', respectively, in the translation).

442. 250a7-8. The word 'weight' is not in Aristotle.

443. Although the Simplicius MSS have 'as A is to F' at this point, it appears that Alexander really read 'as F is to A' (so Diels). For one thing, there is the evidence of 1107,25, which indicates that Alexander's text of the *Physics* had 'so F is to A'. Further, Simplicius himself goes on in lines 4-5 below to suggest that Alexander read 'as F is to A'.

444. This *ê*, though found in *Physics* MSS HI, is not in the text Ross prints; including it makes an already difficult sentence even harder to understand. Citing 1107,2.10, Ross (p. 685) argues that despite its presence here in some Simplicius MSS Simplicius himself did not read it (Simplicius MS M omits it).

445. Simplicius seems to be taking exception to Alexander's reference in 1108,2-3 to double the length C. What Simplicius says here is consistent with his own explanation of the phrase from *Physics* 250a14-15 (see 1107,2-12 above).

446. *logos* here could mean 'claim' or 'statement', but in lines 19 and 28 below Simplicius seems to treat it as meaning 'argument', and accordingly it is translated that way here. When Aristotle says that Zeno's argument is not 'true', he means that the conclusion is false (see *An. Post.* 1.32, 88a18-26; cf. *SE* 1.18, 176b31-177a6).

447. The second part of the lemma omits *analogon gar* ('for [it is] proportional'), which appears in 250a28 as the final words of the *Physics* passage. However, though they are missing from the lemma, Simplicius in fact has these words in 1109,18, and hence there is good reason to think that they stood in his text of the *Physics*.

448. Simplicius uses the verb *eresthai*, which literally means 'to ask' and is part of the technical vocabulary for posing a dialectical question. The verb *erôtan* in line 28 is another verb meaning 'to ask' possessing this technical use.

449. A medimnus is a dry measure equal to about 45 quarts (50 litres).

450. Aristotle's formulation of Zeno's argument, along with the present passage from Simplicius' commentary, make up section A29 in the testimonies concerning Zeno's life and teachings in Diels-Kranz (1:254-5). Aristotle's attribution guarantees that the argument is not simply a Hellenistic or late antique fabrication, but we do not know its source. It could have been part of the Eleatic critique of sense perception. In any case, there seems to be no chance that the dialogue version presented here goes back to Zeno himself, despite the fact that in Diogenes Laertius (3,48) some people are said to have credited Zeno with being the first to write dialogues. (On that claim see section A14 in Diels-Kranz (1:250-51); however, it should be mentioned that Ross, in his Oxford Classical Texts edition, follows Waitz in bracketing Zeno's name in *SE* 170b23.) The critical apparatus in Diels-Kranz indicates that Diels made the suggestion that the version which Simplicius gives might have come from a dialogue by Alkidamas, a student of Gorgias. For a good treatment of recent discussions of the philosophical aspects of the argument, see Wardy, pp. 319-27. Wardy has no doubts that Simplicius is not the author: 'Clearly Simplicius himself could never have fabricated the passage, since its vivid, natural writing stands out in sharp contrast with his own wooden, scholastic style' (p. 320 n. 30).

451. 250a22-3. These same words (*tosouton morion*) are also quoted in lines 2-3 above and line 8 below, though in the first case the phrase has been translated simply as 'the portion'.

452. Aristotle discusses cases, including ship-hauling, involving similar threshold phenomena in *Phys*. 8.3, 253b14-26 (see 1196,32-1200,9 in Simplicius' commentary).

453. Since both human beings and millet seeds (though, perhaps, not their parts) exist actually even when they are part of a larger group, this approach has its limitations. See lines 25-9 below.

454. In this sentence Simplicius is quoting and paraphrasing 250a25-7.

455. 250a28 (missing from the lemma). Simplicius adds *estin* ('it is'), which otherwise would be understood.

456. Simplicius gives the masculine as well as feminine forms of both the definite article and also the Greek word for 'one' because the Zeno of the dialogue treats *kenkhros* ('millet seed') as masculine whereas Aristotle treats it as feminine.

457. Diels punctuates this as a question, and so understood it could be translated: 'Where might I stand and move the earth?' However, Simplicius treats it as a boast, and hence it seems better to use the customary version from Pappus *Synagoge* viii, 11, 19 (Hultsch 1060,1-4).

458. See 1106,24-6 and 1109,21-3 above.

459. This last qualification is added because an incorporeal mover can have infinite power (cf. 1328,23-5).

460. Literally, 'the sound departs from the magnitude' (*apoleipei to megethos ho psophos*).

461. This could also be 'the other [motions]', and then the point would be that the other motions are the same in the relevant respects. In the summary of this passage at the end of the book, Simplicius uses the word *arkhai* ('principles', 1116,9), and it is conceivable that that word has dropped out here (or that we should read *arkhôn* here in line 3 instead of *allôn*).

462. The Greek word is *analogia*, usually translated 'proportion'.

463. In this passage 'power' seems a more appropriate translation of *dunamis* than 'force'.

464. Simplicius is reminding us that, despite Aristotle's use of *poson* in 250a31, the proper terms of comparison for quality are 'more' (*mallon*) and 'less' (*hêtton*). See 249b25-6 and the Subject Index s.v. 'quality' and 'similarity'.

465. Ross (p. 686) argues that *to diplasion* in 250b2 refers to double the amount of change. However, Wardy (p. 329 n. 37) seems to be right when he maintains against Ross that it refers to double the thing being altered, even if the claim is later retracted.

466. 250b3. Simplicius has *en tôi hêmisei to hêmisu* whereas the *Physics* has *en hêmisei hêmisu*. There is no problem about Simplicius' *en tôi hêmisei*, which refers to the time and does not differ in any important way from the *Physics*' *en hêmisei*. However, if *to hêmisu* is interpreted as referring to what is altered, in conformity with what appears to be the convention about use of the definite article in the *Physics* passage, Simplicius would seem to be saying that half the thing is altered in half the time. But then this clause in Simplicius simply repeats the previous one. Accordingly, it seems better take Simplicius' *to hêmisu* as referring to half the amount of alteration rather than to the thing being altered. In fact, *to hêmisu* for *hêmisu* may be just a slip. Though it shows up in the quotation here, there is no evidence that *to hêmisu* ever stood in this place in the *Physics*. Also, it is easy to see how the run of the text (which does have *to hêmisu* earlier in 1111,15, correctly taken from 250b2-3) might have induced a copyist (or even Simplicius) to insert the *to* by mistake.

467. Simplicius appears to take *en isôi diplasion* (literally, 'in an equal, double') in 250b3-4 as involving doubling the alteration in a given time by doubling the original power and keeping the original thing being altered. In the *Physics*, however, this case seems to involve doubling the alteration by keeping the original power and halving the thing being altered.

468. This and the immediately following reference to half the power are not present in the corresponding place in the *Physics*. Simplicius may be troubled by the two references to *half* the time in 250b5-6 where one might expect references to *double* the time. By halving the power as well he may be seeking to bring what Aristotle says here more into line with 250a12-15. Ross (pp. 686-7) suggests that Aristotle may be making the points that the whole force may in fact not move half the weight in half the time and that the motion or change, if and when it comes, may occur all at once. This interpretation, endorsed by Wardy (pp. 329-30), does offer a possible solution.

469. The presence of *epi toutôn* ('in the case of these') in line 25 (and in Simplicius' summary in 1116,14) suggests that we might also read that same phrase here in lines 26-7 instead of *epi tautês* ('in the case of this').

470. cf. 1037,9 above.

471. The Subject Index can be consulted for references to the original discussions.

472. See the footnote on 1039,11 above.

473. In fact, this point is not proved in *Physics* 6.7, where Aristotle deals with such topics. However, Aristotle does prove in *Phys*. 6.7, 238a20-b22 that in a finite time there can be neither an infinite motion nor the motion of an infinite magnitude.

474. Here *pathos*, instead of referring to the quality involved as in the earlier discussion (see, e.g. 1099,7-8), seems to refer to the alteration itself. For recognition of this meaning, see 1295,2-4 and *in Cat*. 208,12-14; 260,9-33; 331,27-332,5 (where the entry on *pathos* in *Metaph*. 5.21, 1022b15-21 is quoted nearly in full). However, in line 33 below *pathos* is once again used for the quality.

475. Diels follows the Aldine edition in inserting *kai* in line 14, but the wording in lines 12-16 is very close to 1103,3-7 and comparison with 1103,6 makes it clear that it is *hoion* which has been left out.

476. This should be case 2, but there appears to be an error here, since the distance is supposed to be the whole original distance. Cf. *Physics* 250a3 and what Simplicius says in 1104,6-11 (which he seems to be following closely here) and also in 1103,13-15.

Bibliography

Brown, V., Review of Rescher and Marmura, *The Refutation by Alexander of Aphrodisias of Galen's Treatise on the Theory of Motion*, *Journal of the Royal Asiatic Society*, 1972, 152-4.

Charlton, W., *Aristotle's 'Physics': Books I and II*, Oxford 1970.

Cornford, F.M., 'Aristotle, *Physics* 250A9-19 and 266A12-24', *Classical Quarterly* 26, 1932, 52-4.

Diels, H., ed., *Die Fragmente der Vorsokratiker*, rev. by W. Kranz, 6th ed., 3 vols., Berlin 1951-52.

——, 'Zur Textgeschichte der Aristotelischen Physik', in *Abhandlungen der Königlichen Preussischen Akademie der Wissenschaften zu Berlin, Phil.-hist. Kl. I*, 1882, 1-42; also in H. Diels, *Kleine Schriften zur Geschichte der antiken Philosophie*, Walter Burkert, ed., Hildesheim 1969, 199-238.

Ess, J. van, Review of Rescher and Marmura, *The Refutation by Alexander of Aphrodisias of Galen's Treatise on the Theory of Motion*, *Erasmus* 24, 1972, 580-3.

Hadot, I., 'La vie et l'oeuvre de Simplicius d'après des sources grecques et arabes', in id., ed., *Simplicius, sa vie, son oeuvre, sa survie* (= Peripatoi 15), Berlin-New York 1987, 3-39; also translated in a slightly revised version as 'The life and work of Simplicius in Greek and Arabic sources' in Sorabji, ed., *Aristotle Transformed*, 275-303.

Jaeger, W., *Aristotle: Fundamentals of the History of His Development*, R. Robinson, trans., 2nd ed., Oxford 1962.

Konstan, D., trans., *Simplicius on Aristotle Physics 6*, London-Ithaca NY 1989.

Owen, G.E.L., 'Logic and metaphysics in some earlier works of Aristotle', in I. Düring and G.E.L. Owen, eds, *Aristotle and Plato in the Mid-Fourth Century* (= Studia Graeca et Latina Gothoburgensia 11), Göteborg 1960; also in G.E.L. Owen, *Logic, Science and Dialectic*, M. Nussbaum, ed., London 1986, 180-99.

Manuwald, B., *Das Buch H der aristotelischen 'Physik': eine Untersuchung zur Einheit und Echtheit* (= Beiträge zur klassischen Philologie 36), Meisenheim am Glan 1971.

Pinès, S., 'Omne quod movetur necesse est ab aliquo moveri: a refutation of Galen by Alexander of Aphrodisias and the theory of motion', *Isis* 52, 1961, 21-54.

Radl, A., *Der Magnetstein in der Antike: Quellen und Zusammenhänge* (= Boethius: Texte und Abhandlungen zur Geschichte der exacten Wissenschaften 19), Stuttgart 1988.

Rescher, N., and Marmura, M.E., eds, *The Refutation by Alexander of Aphrodisias of Galen's Treatise on the Theory of Motion*, Islamabad 1965.

Ross, W.D., *Aristotle's 'Physics': A Revised Text with Introduction and Commentary*, Oxford 1936 (rpr. 1979).

Sharples, R.W., 'The School of Alexander?' in Sorabji, ed., *Aristotle Transformed*, 83-111.

——, 'Alexander of Aphrodisias: scholasticism and innovation', in W. Haase, ed., *Aufstieg und Niedergang der Römischen Welt*, II, 36.2, Berlin-New York 1987, 1176-1243.

Sorabji, R., ed., *Aristotle Transformed: The Ancient Commentators and Their Influence*, London-Ithaca NY 1990.

——, *Matter, Space and Motion: Theories in Antiquity and Their Sequel*, London-Ithaca NY 1988.

——, ed., *Philoponus and the Rejection of Aristotelian Science*, London-Ithaca NY 1987.

Tarán, L., 'The text of Simplicius' commentary on Aristotle's *Physics*', in I. Hadot, ed., *Simplicius, sa vie, son oeuvre, sa survie* (= Peripatoi 15), Berlin-New York 1987, 246-66.

——, Review of P. Moraux, *Du Ciel*, *Gnomon* 46, 1974, 121-42.

Verbeke, G., 'L'argument du livre VII de la *Physique*: une impasse philosophique', in I. Düring, ed., *Naturphilosophie bei Aristoteles und Theophrast: Verhandlungen des 4. symposium Aristotelicum*, Heidelberg 1969, 250-67.

Wagner, H., 'Über den Charakter des VII. Buches des aristotelischen Physikvorlesung', *Archiv für Geschichte der Philosophie* 56, 1974, 172-81.

Wallies, M., 'Zur doppelten Rezension des siebenten Buches der aristotelischen Physik', *Rheinisches Museum für Philologie* 70, 1915, 147-9.

Wardy, R., *The Chain of Change: A Study of Aristotle's 'Physics' VII*, Cambridge 1990.

Wehrli, F., ed., *Die Schule des Aristoteles*, vol. 8: *Eudemos von Rhodos*, 2nd ed., Basel-Stuttgart 1969.

Wicksteed, P.H. and Cornford, F.M., trans., *The Physics with an English Translation* (= Loeb edition), vol. 2, London-Cambridge Mass. 1934.

Philosophers Cited by Simplicius

Alexander (of Aphrodisias). Peripatetic philosopher and teacher. Appointed to Peripatetic chair, possibly in Athens, about AD 200. Author of commentaries on Aristotle and of other treatises (though not everything transmitted under his name is genuine). Simplicius quotes extensively from his lost commentary on the *Physics*.

Anaxagoras. Fifth-century BC Presocratic Pluralist. According to him, the universe began from a state in which 'all things were together'. Motion and separation were introduced into this originally uniform mixture by *Nous* (Intellect or Mind).

Archimedes. Third-century BC Syracusan mathematician, engineer and inventor. Simplicius refers to his weighing machine and his boast about moving the earth.

Aristotle. Fourth-century BC philosopher, scientist and teacher. Studied at Plato's Academy for twenty years. Away from Athens for approximately twelve years after Plato's death, but later returned and instituted the Lyceum as an educational and research institution. Founder of Peripatetic school.

Democritus. Fifth-century BC Presocratic. Developed Leucippus' view that physical things are collections of minute, invisible atoms separated by void. Prolific author on a wide range of topics, but most of what he wrote has perished.

Eudemus (of Rhodes). Fourth-century BC Peripatetic. Pupil and associate of Aristotle. Author of histories of arithmetic, geometry, astronomy, and theology. Simplicius refers to his *Physics*, which largely followed the organization and content of Aristotle's *Physics* (except for book 7, which he ignored).

Galen. Second-century AD physician from Pergamum. His voluminous writings deal not only with medical topics but also with philosophical views.

Leucippus. Fifth-century BC Presocratic. Originator of Atomism, further developed by his pupil Democritus (q.v.).

Peripatetics. Followers of Aristotle.

Plato. Lived *c.* 428-348 BC. Athenian philosopher and author of dialogues inspired in part by the teachings of Socrates and the Pythagoreans. Founder of the Academy, teacher of Aristotle.

Plotinus. Third-century AD Neoplatonist from Egypt who lived and taught in Rome. Author of the *Enneads*, edited by his pupil Porphyry.

Protagoras. Fifth-century BC Sophist from Abdera. Often portrayed (e.g. by Plato in the *Theaetetus*) as a relativist and a champion of sense perception.

Pythagoreans. Followers of Pythagoras, who himself died about 500 BC. The views of the school he founded were influential for centuries and played a role in the development of Platonism and, later on, of Neoplatonism. After a revival in the first century BC, there was a large Pythagorean literature, some of it falsely claiming to be the work of early members of the school.

Themistius. Fourth-century AD rhetorician and philosopher known for his paraphrases of Aristotle's works, notably the *Physics*. Other exegetical works on Aristotle and Plato have not survived.

Zeno. Fifth-century BC Presocratic. Follower of the Eleatic Parmenides. Propounder of paradoxes intended to undermine reliance on sense perception and to refute common-sense beliefs in plurality and motion.

Appendix
The Commentators*

The 15,000 pages of the Ancient Greek Commentaries on Aristotle are the largest corpus of Ancient Greek philosophy that has not been translated into English or other European languages. The standard edition (*Commentaria in Aristotelem Graeca*, or *CAG*) was produced by Hermann Diels as general editor under the auspices of the Prussian Academy in Berlin. Arrangements have now been made to translate at least a large proportion of this corpus, along with some other Greek and Latin commentaries not included in the Berlin edition, and some closely related non-commentary works by the commentators.

The works are not just commentaries on Aristotle, although they are invaluable in that capacity too. One of the ways of doing philosophy between A.D. 200 and 600, when the most important items were produced, was by writing commentaries. The works therefore represent the thought of the Peripatetic and Neoplatonist schools, as well as expounding Aristotle. Furthermore, they embed fragments from all periods of Ancient Greek philosophical thought: this is how many of the Presocratic fragments were assembled, for example. Thus they provide a panorama of every period of Ancient Greek philosophy.

The philosophy of the period from A.D. 200 to 600 has not yet been intensively explored by philosophers in English-speaking countries, yet it is full of interest for physics, metaphysics, logic, psychology, ethics and religion. The contrast with the study of the Presocratics is striking. Initially the incomplete Presocratic fragments might well have seemed less promising, but their interest is now widely known, thanks to the philological and philosophical effort that has been concentrated upon them. The incomparably vaster corpus which preserved so many of those fragments offers at least as much interest, but is still relatively little known.

The commentaries represent a missing link in the history of philosophy: the Latin-speaking Middle Ages obtained their knowledge of Aristotle at least partly through the medium of the commentaries. Without an appreciation of this, mediaeval interpretations of Aristotle will not be understood. Again, the ancient commentaries are the unsuspected source of ideas which have been thought, wrongly, to originate in the later mediaeval period. It has been supposed, for example, the Bonaventure in the thirteenth century invented the ingenious arguments based on the concept of infinity which attempt to prove the Christian view that the universe had a beginning. In fact, Bonaventure is merely repeating arguments devised

* Reprinted from the Editor's General Introduction to the series in Christian Wildberg, *Philoponus Against Aristotle on the Eternity of the World*, London and Ithaca, N.Y., 1987.

by the commentator Philoponus 700 years earlier and preserved in the meantime by the Arabs. Bonaventure even uses Philoponus' original examples. Again, the introduction of impetus theory into dynamics, which has been called a scientific revolution, has been held to be an independent invention of the Latin West, even if it was earlier discovered by the Arabs or their predecessors. But recent work has traced a plausible route by which it could have passed from Philoponus, via the Arabs, to the West.

The new availability of the commentaries in the sixteenth century, thanks to printing and to fresh Latin translations, helped to fuel the Renaissance break from Aristotelian science. For the commentators record not only Aristotle's theories, but also rival ones, while Philoponus as a Christian devises rival theories of his own and accordingly is mentioned in Galileo's early works more frequently than Plato.[1]

It is not only for their philosophy that the works are of interest. Historians will find information about the history of schools, their methods of teaching and writing and the practices of an oral tradition.[2] Linguists will find the indexes and translations an aid for studying the development of word meanings, almost wholly uncharted in Liddell and Scott's *Lexicon*, and for checking shifts in grammatical usage.

Given the wide range of interests to which the volumes will appeal, the aim is to produce readable translations, and to avoid so far as possible presupposing any knowledge of Greek. Notes will explain points of meaning, give cross-references to other works, and suggest alternative interpretations of the text where the translator does not have a clear preference. The introduction to each volume will include an explanation why the work was chosen for translation: none will be chosen simply because it is there. Two of the Greek texts are currently being re-edited – those of Simplicius *in Physica* and *in de Caelo* – and new readings will be exploited by

1 See Fritz Zimmermann, 'Philoponus' impetus theory in the Arabic tradition'; Charles Schmitt, 'Philoponus' commentary on Aristotle's *Physics* in the sixteenth century', and Richard Sorabji, 'John Philoponus', in Richard Sorabji (ed.), *Philoponus and the Rejection of Aristotelian Science* (London and Ithaca, N.Y. 1987).

2 See e.g. Karl Praechter, 'Die griechischen Aristoteleskommentare', *Byzantinische Zeitschrift* 18 (1909), 516-38 (translated into English in R. Sorabji (ed.), *Aristotle Transformed: the ancient commentators and their influence* (London and Ithaca, N.Y. 1990); M. Plezia, *de Commentariis Isagogicis* (Cracow 1947); M. Richard, *'Apo Phônês', Byzantion* 20 (1950), 191-222; É. Evrard, *L'Ecole d'Olympiodore et la composition du commentaire à la physique de Jean Philopon*, Diss. (Liège 1957); L.G. Westerink, *Anonymous Prolegomena to Platonic Philosophy* (Amsterdam 1962) (new revised edition, translated into French, Collection Budé; part of the revised introduction, in English, is included in *Aristotle Transformed*); A.-J. Festugière, 'Modes de composition des commentaires de Proclus', *Museum Helveticum* 20 (1963), 77-100, repr. in his *Études* (1971), 551-74; P. Hadot, 'Les divisions des parties de la philosophie dans l'antiquité', *Museum Helveticum* 36 (1979), 201-23; I. Hadot, 'La division néoplatonicienne des écrits d'Aristote', in J. Wiesner (ed.), *Aristoteles Werk und Wirkung* (Paul Moraux gewidmet), vol. 2 (Berlin 1986); I. Hadot, 'Les introductions aux commentaires exégétiques chez les auteurs néoplatoniciens et les auteurs chrétiens', in M. Tardieu (ed.), *Les règles de l'interprétation* (Paris 1987), 99-119. These topics are treated, and a bibliography supplied, in *Aristotle Transformed*.

translators as they become available. Each volume will also contain a list of proposed emendations to the standard text. Indexes will be of more uniform extent as between volumes than is the case with the Berlin edition, and there will be three of them: an English-Greek glossary, a Greek-English index, and a subject index.

The commentaries fall into three main groups. The first group is by authors in the Aristotelian tradition up to the fourth century A.D. This includes the earliest extant commentary, that by Aspasius in the first half of the second century A.D. on the *Nicomachean Ethics*. The anonymous commentary on Books 2, 3, 4 and 5 of the *Nicomachean Ethics*, in *CAG* vol. 20, is derived from Adrastus, a generation later.[3] The commentaries by Alexander of Aphrodisias (appointed to his chair between A.D. 198 and 209) represent the fullest flowering of the Aristotelian tradition. To his successors Alexander was The Commentator *par excellence*. To give but one example (not from a commentary) of his skill at defending and elaborating Aristotle's views, one might refer to his defence of Aristotle's claim that space is finite against the objection that an edge of space is conceptually problematic.[4] Themistius (*fl.* late 340s to 384 or 385) saw himself as the inventor of paraphrase, wrongly thinking that the job of commentary was completed.[5] In fact, the Neoplatonists were to introduce new dimensions into commentary. Themistius' own relation to the Neoplatonist as opposed to the Aristotelian tradition is a matter of controversy,[6] but it would be agreed that his commentaries show far less bias than the full-blown Neoplatonist ones. They are also far more informative than the designation 'paraphrase' might suggest, and it has been estimated that Philoponus' *Physics* commentary draws silently on Themistius six hundred times.[7] The pseudo-Alexandrian commentary on *Metaphysics* 6-14, of unknown

3 Anthony Kenny, *The Aristotelian Ethics* (Oxford 1978), 37, n. 3: Paul Moraux, *Der Aristotelismus bei den Griechen*, vol. 2 (Berlin 1984), 323-30.

4 Alexander, *Quaestiones* 3.12, discussed in my *Matter, Space and Motion* (London and Ithaca, N.Y. 1988). For Alexander see R.W. Sharples, 'Alexander of Aphrodisias: scholasticism and innovation', in W. Haase (ed.), *Aufstieg und Niedergang der römischen Welt*, part 2 *Principat*, vol. 36.2, *Philosophie und Wissenschaften* (1987).

5 Themistius *in An. Post.* 1,2-12. See H.J. Blumenthal, 'Photius on Themistius (Cod. 74): did Themistius write commentaries on Aristotle?', *Hermes* 107 (1979), 168-82.

6 For different views, see H.J. Blumenthal, 'Themistius, the last Peripatetic commentator on Aristotle?', in Glen W. Bowersock, Walter Burkert, Michael C.J. Putnam, *Arktouros*, Hellenic Studies Presented to Bernard M.W. Knox (Berlin and N.Y., 1979), 391-400; E.P. Mahoney, 'Themistius and the agent intellect in James of Viterbo and other thirteenth-century philosophers: (Saint Thomas Aquinas, Siger of Brabant and Henry Bate)', *Augustiniana* 23 (1973), 422-67, at 428-31; id., 'Neoplatonism, the Greek commentators and Renaissance Aristotelianism', in D.J. O'Meara (ed.), *Neoplatonism and Christian Thought* (Albany N.Y. 1982), 169-77 and 264-82, esp. n. 1, 264-6; Robert Todd, introduction to translation of Themistius *in DA* 3.4-8, in *Two Greek Aristotelian Commentators on the Intellect*, trans. Frederick M. Schroeder and Robert B. Todd (Toronto 1990).

7 H. Vitelli, *CAG* 17, p. 992, s.v. Themistius.

authorship, has been placed by some in the same group of commentaries as being earlier than the fifth century.[8]

By far the largest group of extant commentaries is that of the Neoplatonists up to the sixth century A.D. Nearly all the major Neoplatonists, apart from Plotinus (the founder of Neoplatonism), wrote commentaries on Aristotle, although those of Iamblichus (*c.* 250–*c.* 325) survive only in fragments, and those of three Athenians, Plutarchus (died 432), his pupil Proclus (410–485) and the Athenian Damascius (*c.* 462–after 538), are lost.[9] As a result of these losses, most of the extant Neoplatonist commentaries come from the late fifth and the sixth centuries and a good proportion from Alexandria. There are commentaries by Plotinus' disciple and editor Porphyry (232–309), by Iamblichus' pupil Dexippus (*c.* 330), by Proclus' teacher Syrianus (died *c.* 437), by Proclus' pupil Ammonius (435/445–517/526), by Ammonius' three pupils Philoponus (*c.* 490 to 570s), Simplicius (wrote after 532, probably after 538) and Asclepius (sixth century), by Ammonius' next but one successor Olympiodorus (495/505–after 565), by Elias (*fl.* 541?), by David (second half of the sixth century, or beginning of the seventh) and by Stephanus (took the chair in Constantinople *c.* 610). Further, a commentary on the *Nicomachean Ethics* has been ascribed to Heliodorus of Prusa, an unknown pre-fourteenth-century figure, and there is a commentary by Simplicius' colleague Priscian of Lydia on Aristotle's successor Theophrastus. Of these commentators some of the last were Christians (Philoponus, Elias, David and Stephanus), but they were Christians writing in the Neoplatonist tradition, as was also Boethius who produced a number of commentaries in Latin before his death in 525 or 526.

The third group comes from a much later period in Byzantium. The Berlin edition includes only three out of more than a dozen commentators described in Hunger's *Byzantinisches Handbuch*.[10] The two most important are Eustratius (1050/1060–*c.*1120), and Michael of Ephesus. It has been suggested that these two belong to a circle organised by the princess

8 The similarities to Syrianus (died *c.* 437) have suggested to some that it predates Syrianus (most recently Leonardo Tarán, review of Paul Moraux, *Der Aristotelismus*, vol.1 in *Gnomon* 46 (1981), 721-50 at 750), to others that it draws on him (most recently P. Thillet, in the Budé edition of Alexander *de Fato*, p. lvii). Praechter ascribed it to Michael of Ephesus (eleventh or twelfth century), in his review of *CAG* 22.2, in *Göttingische Gelehrte Anzeiger* 168 (1906), 861-907.

9 The Iamblichus fragments are collected in Greek by Bent Dalsgaard Larsen, *Jamblique de Chalcis, Exégète et Philosophe* (Aarhus 1972), vol. 2. Most are taken from Simplicius, and will accordingly be translated in due course. The evidence on Damascius' commentaries is given in L.G. Westerink, *The Greek Commentaries on Plato's Phaedo*, vol. 2, Damascius (Amsterdam 1977), 11-12; on Proclus' in L.G. Westerink, *Anonymous Prolegomena to Platonic Philosophy* (Amsterdam 1962), xii, n. 22; on Plutarchus' in H.M. Blumenthal, 'Neoplatonic elements in the de Anima commentaries', *Phronesis* 21 (1976), 75.

10 Herbert Hunger, *Die hochsprachliche profane Literatur der Byzantiner*, vol. 1 (= *Byzantinisches Handbuch*, part 5, vol. 1) (Munich 1978), 25-41. See also B.N. Tatakis, *La Philosophie Byzantine* (Paris 1949).

Anna Comnena in the twelfth century, and accordingly the completion of Michael's commentaries has been redated from 1040 to 1138.[11] His commentaries include areas where gaps had been left. Not all of these gap-fillers are extant, but we have commentaries on the neglected biological works, on the *Sophistici Elenchi*, and a small fragment of one on the *Politics*. The lost *Rhetoric* commentary had a few antecedents, but the *Rhetoric* too had been comparatively neglected. Another product of this period may have been the composite commentary on the *Nicomachean Ethics* (*CAG* 20) by various hands, including Eustratius and Michael, along with some earlier commentators, and an improvisation for Book 7. Whereas Michael follows Alexander and the conventional Aristotelian tradition, Eustratius' commentary introduces Platonist, Christian and anti- Islamic elements.[12]

The composite commentary was to be translated into Latin in the next century by Robert Grosseteste in England. But Latin translations of various logical commentaries were made from the Greek still earlier by James of Venice (*fl. c.* 1130), a contemporary of Michael of Ephesus, who may have known him in Constantinople. And later in that century other commentaries and works by commentators were being translated from Arabic versions by Gerard of Cremona (died 1187).[13] So the twelfth century resumed the transmission which had been interrupted at Boethius' death in the sixth century.

The Neoplatonist commentaries of the main group were initiated by Porphyry. His master Plotinus had discussed Aristotle, but in a very independent way, devoting three whole treatises (*Enneads* 6.1-3) to attacking Aristotle's classification of the things in the universe into categories. These categories took no account of Plato's world of Ideas, were inferior to Plato's classifications in the *Sophist* and could anyhow be collapsed, some

11 R. Browning, 'An unpublished funeral oration on Anna Comnena', *Proceedings of the Cambridge Philological Society* n.s. 8 (1962), 1-12, esp. 6-7.

12 R. Browning, op. cit. H.D.P. Mercken, *The Greek Commentaries of the Nicomachean Ethics of Aristotle in the Latin Translation of Grosseteste, Corpus Latinum Commentariorum in Aristotelem Graecorum* VI 1 (Leiden 1973), ch. 1, 'The compilation of Greek commentaries on Aristotle's Nicomachean Ethics'. Sten Ebbesen, 'Anonymi Aurelianensis I Commentarium in *Sophisticos Elenchos*', *Cahiers de l'Institut Moyen Age Grecque et Latin* 34 (1979), 'Boethius, Jacobus Veneticus, Michael Ephesius and "Alexander" ', pp. v-xiii; id., *Commentators and Commentaries on Aristotle's Sophistici Elenchi*, 3 parts, *Corpus Latinum Commentariorum in Aristotelem Graecorum*, vol. 7 (Leiden 1981); A. Preus, *Aristotle and Michael of Ephesus on the Movement and Progression of Animals* (Hildesheim 1981), introduction.

13 For Grosseteste, see Mercken as in n. 12. For James of Venice, see Ebbesen as in n. 12, and L. Minio-Paluello, 'Jacobus Veneticus Grecus', *Traditio* 8 (1952), 265-304; id., 'Giacomo Veneto e l'Aristotelismo Latino', in Pertusi (ed.), *Venezia e l'Oriente fra tardo Medioevo e Rinascimento* (Florence 1966), 53-74, both reprinted in his *Opuscula* (1972). For Gerard of Cremona, see M. Steinschneider, *Die europäischen Übersetzungen aus dem arabischen bis Mitte des 17. Jahrhunderts* (repr. Graz 1956); E. Gilson, *History of Christian Philosophy in the Middle Ages* (London 1955), 235-6 and more generally 181-246. For the translators in general, see Bernard G. Dod, 'Aristoteles Latinus', in N. Kretzmann, A. Kenny, J. Pinborg (eds), *The Cambridge History of Latin Medieval Philosophy* (Cambridge 1982).

of them into others. Porphyry replied that Aristotle's categories could apply perfectly well to the world of intelligibles and he took them as in general defensible.[14] He wrote two commentaries on the *Categories*, one lost, and an introduction to it, the *Isagôgê*, as well as commentaries, now lost, on a number of other Aristotelian works. This proved decisive in making Aristotle a necessary subject for Neoplatonist lectures and commentary. Proclus, who was an exceptionally quick student, is said to have taken two years over his Aristotle studies, which were called the Lesser Mysteries, and which preceded the Greater Mysteries of Plato.[15] By the time of Ammonius, the commentaries reflect a teaching curriculum which begins with Porphyry's *Isagôgê* and Aristotle's *Categories*, and is explicitly said to have as its final goal a (mystical) ascent to the supreme Neoplatonist deity, the One.[16] The curriculum would have progressed from Aristotle to Plato, and would have culminated in Plato's *Timaeus* and *Parmenides*. The latter was read as being about the One, and both works were established in this place in the curriculum at least by the time of Iamblichus, if not earlier.[17]

Before Porphyry, it had been undecided how far a Platonist should accept Aristotle's scheme of categories. But now the proposition began to gain force that there was a harmony between Plato and Aristotle on most things.[18] Not for the only time in the history of philosophy, a perfectly crazy proposition proved philosophically fruitful. The views of Plato and of Aristotle had both to be transmuted into a new Neoplatonist philosophy in order to exhibit the supposed harmony. Iamblichus denied that Aristotle contradicted Plato on the theory of Ideas.[19] This was too much for Syrianus and his pupil Proclus. While accepting harmony in many areas,[20] they could see that there was disagreement on this issue and also on the issue of whether God was causally responsible for the existence of the ordered

14 See P. Hadot, 'L'harmonie des philosophies de Plotin et d'Aristote selon Porphyre dans le commentaire de Dexippe sur les Catégories', in *Plotino e il neoplatonismo in Oriente e in Occidente* (Rome 1974), 31-47; A.C. Lloyd, 'Neoplatonic logic and Aristotelian logic', *Phronesis* 1 (1955-6), 58-79 and 146-60.

15 Marinus, *Life of Proclus* ch. 13, 157,41 (Boissonade).

16 The introductions to the *Isagôgê* by Ammonius, Elias and David, and to the *Categories* by Ammonius, Simplicius, Philoponus, Olympiodorus and Elias are discussed by L.G. Westerink, *Anonymous Prolegomena* and I. Hadot, 'Les Introductions', see n. 2 above.

17 Proclus in *Alcibiadem 1* p. 11 (Creuzer); Westerink, *Anonymous Prolegomena*, ch. 26, 12f. For the Neoplatonist curriculum see Westerink, Festugière, P. Hadot and I. Hadot in n. 2.

18 See e.g. P. Hadot (1974), as in n. 14 above; H.J. Blumenthal, 'Neoplatonic elements in the de Anima commentaries', *Phronesis* 21 (1976), 64-87; H.A. Davidson, 'The principle that a finite body can contain only finite power', in S. Stein and R. Loewe (eds), *Studies in Jewish Religious and Intellectual History presented to A. Altmann* (Alabama 1979), 75-92; Carlos Steel, 'Proclus et Aristotle', Proceedings of the Congrès Proclus held in Paris 1985, J. Pépin and H.D. Saffrey (eds), *Proclus, lecteur et interprète des anciens* (Paris 1987), 213-25; Koenraad Verrycken, *God en Wereld in de Wijsbegeerte van Ioannes Philoponus*, Ph.D. Diss. (Louvain 1985).

19 Iamblichus ap. Elian *in Cat.* 123,1-3.

20 Syrianus *in Metaph.* 80,4-7; Proclus *in Tim.* 1.6,21-7,16.

physical cosmos, which Aristotle denied. But even on these issues, Proclus' pupil Ammonius was to claim harmony, and, though the debate was not clear cut,[21] his claim was on the whole to prevail. Aristotle, he maintained, accepted Plato's Ideas,[22] at least in the form of principles (*logoi*) in the divine intellect, and these principles were in turn causally responsible for the beginningless existence of the physical universe. Ammonius wrote a whole book to show that Aristotle's God was thus an efficent cause, and though the book is lost, some of its principal arguments are preserved by Simplicius.[23] This tradition helped to make it possible for Aquinas to claim Aristotle's God as a Creator, albeit not in the sense of giving the universe a beginning, but in the sense of being causally responsible for its beginningless existence.[24] Thus what started as a desire to harmonise Aristotle with Plato finished by making Aristotle safe for Christianity. In Simplicius, who goes further than anyone,[25] it is a formally stated duty of the commentator to display the harmony of Plato and Aristotle in most things.[26] Philoponus, who with his independent mind had thought better of his earlier belief in harmony, is castigated by Simplicius for neglecting this duty.[27]

The idea of harmony was extended beyond Plato and Aristotle to Plato and the Presocratics. Plato's pupils Speusippus and Xenocrates saw Plato as being in the Pythagorean tradition.[28] From the third to first centuries B.C., pseudo-Pythagorean writings present Platonic and Aristotelian doctrines as if they were the ideas of Pythagoras and his pupils,[29] and these forgeries were later taken by the Neoplatonists as genuine. Plotinus saw the Presocratics as precursors of his own views,[30] but Iamblichus went far beyond him by writing ten volumes on Pythagorean philosophy.[31] Thereafter Proclus sought to unify the whole of

21 Asclepius sometimes accepts Syranius' interpretation (*in Metaph.* 433,9-436,6); which is, however, qualified, since Syrianus thinks Aristotle is really committed willy-nilly to much of Plato's view (*in Metaph.* 117,25-118,11; ap. Asclepium *in Metaph.* 433,16; 450,22); Philoponus repents of his early claim that Plato is not the target of Aristotle's attack, and accepts that Plato is rightly attacked for treating ideas as independent entities outside the divine Intellect (*in DA* 37,18-31; *in Phys.* 225,4-226,11; *contra Procl.* 26,24-32,13; *in An. Post.* 242,14-243,25).

22 Asclepius *in Metaph.* from the voice of (i.e. from the lectures of) Ammonius 69,17-21; 71,28; cf. Zacharias *Ammonius, Patrologia Graeca* vol. 85 col. 952 (Colonna).

23 Simplicius *in Phys.* 1361,11-1363,12. See H.A. Davidson; Carlos Steel; Koenraad Verrycken in n. 18 above.

24 See Richard Sorabji, *Matter, Space and Motion* (London and Ithaca N.Y. 1988), ch. 15.

25 See e.g. H.J. Blumenthal in n. 18 above.

26 Simplicius *in Cat.* 7,23-32.

27 Simplicius *in Cael.* 84,11-14; 159,2-9. On Philoponus' *volte face* see n. 21 above.

28 See e.g. Walter Burkert, *Weisheit und Wissenschaft* (Nürnberg 1962), translated as *Lore and Science in Ancient Pythagoreanism* (Cambridge Mass. 1972), 83-96.

29 See Holge Thesleff, *An Introduction to the Pythagorean Writings of the Hellenistic Period* (Åbo 1961); Thomas Alexander Szlezák, *Pseudo-Archytas über die Kategorien*, Peripatoi vol. 4 (Berlin and New York 1972).

30 Plotinus e.g. 4.8.1; 5.1.8 (10-27); 5.1.9.

31 See Dominic O'Meara, *Pythagoras Revived: Mathematics and Philosophy in Late Antiquity* (Oxford 1989).

Greek philosophy by presenting it as a continuous clarification of divine revelation[32] and Simplicius argued for the same general unity in order to rebut Christian charges of contradictions in pagan philosophy.[33]

Later Neoplatonist commentaries tend to reflect their origin in a teaching curriculum:[34] from the time of Philoponus, the discussion is often divided up into lectures, which are subdivided into studies of doctrine and of text. A general account of Aristotle's philosophy is prefixed to the *Categories* commentaries and divided, according to a formula of Proclus,[35] into ten questions. It is here that commentators explain the eventual purpose of studying Aristotle (ascent to the One) and state (if they do) the requirement of displaying the harmony of Plato and Aristotle. After the ten-point introduction to Aristotle, the *Categories* is given a six-point introduction, whose antecedents go back earlier than Neoplatonism, and which requires the commentator to find a unitary theme or scope (*skopos*) for the treatise. The arrangements for late commentaries on Plato are similar. Since the Plato commentaries form part of a single curriculum they should be studied alongside those on Aristotle. Here the situation is easier, not only because the extant corpus is very much smaller, but also because it has been comparatively well served by French and English translators.[36]

Given the theological motive of the curriculum and the pressure to harmonise Plato with Aristotle, it can be seen how these commentaries are a major source for Neoplatonist ideas. This in turn means that it is not safe to extract from them the fragments of the Presocratics, or of other authors, without making allowance for the Neoplatonist background against which the fragments were originally selected for discussion. For different reasons, analogous warnings apply to fragments preserved by the pre-Neoplatonist commentator Alexander.[37] It will be another advantage of the present translations that they will make it easier to check the distorting effect of a commentator's background.

Although the Neoplatonist commentators conflate the views of Aristotle with those of Neoplatonism, Philoponus alludes to a certain convention

32 See Christian Guérard, 'Parménide d'Elée selon les Néoplatoniciens', forthcoming.

33 Simplicius *in Phys*. 28,32-29,5; 640,12-18. Such thinkers as Epicurus and the Sceptics, however, were not subject to harmonisation.

34 See the literature in n. 2 above.

35 ap. Elian *in Cat*. 107,24-6.

36 English: Calcidius *in Tim*. (parts by van Winden; den Boeft); Iamblichus fragments (Dillon); Proclus *in Tim*. (Thomas Taylor); Proclus *in Parm*. (Dillon); Proclus *in Parm*., end of 7th book, from the Latin (Klibansky, Labowsky, Anscombe); Proclus *in Alcib. 1* (O'Neill); Olympiodorus and Damascius *in Phaedonem* (Westerink); Damascius *in Philebum* (Westerink); *Anonymous Prolegomena to Platonic Philosophy* (Westerink). See also extracts in Thomas Taylor, *The Works of Plato*, 5 vols. (1804). French: Proclus *in Tim*. and *in Rempublicam* (Festugière); *in Parm*. (Chaignet); Anon. *in Parm* (P. Hadot); Damascius *in Parm*. (Chaignet).

37 For Alexander's treatment of the Stoics, see Robert B. Todd, *Alexander of Aphrodisias on Stoic Physics* (Leiden 1976), 24-9.

when he quotes Plutarchus expressing disapproval of Alexander for expounding his own philosophical doctrines in a commentary on Aristotle.[38] But this does not stop Philoponus from later inserting into his own commentaries on the *Physics* and *Meteorology* his arguments in favour of the Christian view of Creation. Of course, the commentators also wrote independent works of their own, in which their views are expressed independently of the exegesis of Aristotle. Some of these independent works will be included in the present series of translations.

The distorting Neoplatonist context does not prevent the commentaries from being incomparable guides to Aristotle. The introductions to Aristotle's philosophy insist that commentators must have a minutely detailed knowledge of the entire Aristotelian corpus, and this they certainly have. Commentators are also enjoined neither to accept nor reject what Aristotle says too readily, but to consider it in depth and without partiality. The commentaries draw one's attention to hundreds of phrases, sentences and ideas in Aristotle, which one could easily have passed over, however often one read him. The scholar who makes the right allowance for the distorting context will learn far more about Aristotle than he would be likely to on his own.

The relations of Neoplatonist commentators to the Christians were subtle. Porphyry wrote a treatise explicitly against the Christians in 15 books, but an order to burn it was issued in 448, and later Neoplatonists were more circumspect. Among the last commentators in the main group, we have noted several Christians. Of these the most important were Boethius and Philoponus. It was Boethius' programme to transmit Greek learning to Latin-speakers. By the time of his premature death by execution, he had provided Latin translations of Aristotle's logical works, together with commentaries in Latin but in the Neoplatonist style on Porphyry's *Isagôgê* and on Aristotle's *Categories* and *de Interpretatione*, and interpretations of the *Prior* and *Posterior Analytics*, *Topics* and *Sophistici Elenchi*. The interruption of his work meant that knowledge of Aristotle among Latin-speakers was confined for many centuries to the logical works. Philoponus is important both for his proofs of the Creation and for his progressive replacement of Aristotelian science with rival theories, which were taken up at first by the Arabs and came fully into their own in the West only in the sixteenth century.

Recent work has rejected the idea that in Alexandria the Neoplatonists compromised with Christian monotheism by collapsing the distinction between their two highest deities, the One and the Intellect. Simplicius (who left Alexandria for Athens) and the Alexandrians Ammonius and Asclepius appear to have acknowledged their beliefs quite openly, as later

38 Philoponus *in DA* 21,20-3.

did the Alexandrian Olympiodorus, despite the presence of Christian students in their classes.[39]

The teaching of Simplicius in Athens and that of the whole pagan Neoplatonist school there was stopped by the Christian Emperor Justinian in 529. This was the very year in which the Christian Philoponus in Alexandria issued his proofs of Creation against the earlier Athenian Neoplatonist Proclus. Archaeological evidence has been offered that, after their temporary stay in Ctesiphon (in present-day Iraq), the Athenian Neoplatonists did not return to their house in Athens, and further evidence has been offered that Simplicius went to Harrān (Carrhae), in present-day Turkey near the Iraq border.[40] Wherever he went, his commentaries are a treasurehouse of information about the preceding thousand years of Greek philosophy, information which he painstakingly recorded after the closure in Athens, and which would otherwise have been lost. He had every reason to feel bitter about Christianity, and in fact he sees it and Philoponus, its representative, as irreverent. They deny the divinity of the heavens and prefer the physical relics of dead martyrs.[41] His own commentaries by contrast culminate in devout prayers.

Two collections of articles by various hands have been published, to make the work of the commentators better known. The first is devoted to Philoponus;[42] the second is about the commentators in general, and goes into greater detail on some of the issues briefly mentioned here.[43]

39 For Simplicius, see I. Hadot, *Le Problème du Néoplatonisme Alexandrin: Hiéroclès et Simplicius* (Paris 1978); for Ammonius and Asclepius, Koenraad Verrycken, *God en wereld in de Wijsbegeerte van Ioannes Philoponus*, Ph.D. Diss. (Louvain 1985); for Olympiodorus, L.G. Westerink, *Anonymous Prolegomena to Platonic Philosophy* (Amsterdam 1962).

40 Alison Frantz, 'Pagan philosophers in Christian Athens', *Proceedings of the American Philosophical Society* 119 (1975), 29-38; M. Tardieu, 'Témoins orientaux du *Premier Alcibiade* à Harrān et à Nag 'Hammādi', *Journal Asiatique* 274 (1986); id., 'Les calendriers en usage à Harrān d'après les sources arabes et le commentaire de Simplicius à la *Physique* d'Aristote', in I. Hadot (ed.), *Simplicius, sa vie, son oeuvre, sa survie* (Berlin 1987), 40-57; id., *Coutumes nautiques mésopotamiennes chez Simplicius*, in preparation. The opposing view that Simplicius returned to Athens is most fully argued by Alan Cameron, 'The last day of the Academy at Athens', *Proceedings of the Cambridge Philological Society* 195, n.s. 15 (1969), 7-29.

41 Simplicius *in Cael.* 26,4-7; 70,16-18; 90,1-18; 370,29-371,4. See on his whole attitude Philippe Hoffmann, 'Simplicius' polemics', in Richard Sorabji (ed.), *Philoponus and the Rejection of Aristotelian Science* (London and Ithaca, N.Y. 1987).

42 Richard Sorabji (ed.), *Philoponus and the Rejection of Aristotelian Science* (London and Ithaca, N.Y. 1987).

43 Richard Sorabji (ed.), *Aristotle Transformed: the ancient commentators and their influence* (London and Ithaca, N.Y. 1990). The lists of texts and previous translations of the commentaries included in Wildberg, *Philoponus Against Aristotle on the Eternity of the World* (pp. 12ff.) are not included here. The list of translations should be augmented by: F.L.S. Bridgman, Heliodorus (?) in *Ethica Nicomachea*, London 1807.

I am grateful for comments to Henry Blumenthal, Victor Caston, I. Hadot, Paul Mercken, Alain Segonds, Robert Sharples, Robert Todd, L.G. Westerink and Christian Wildberg.

English-Greek Glossary

The following abbreviations are used in this glossary: adj., adjective; adv., adverb; mid., middle voice; n., noun; vb., verb; pass., passive voice; pf., perfect tense; ptc., participle.

abide: *hupomenein*
able, be: *dunasthai*
absolutely: *haplôs*, *pantôs*
absolutely all, absolutely every: *hapas*
absurd: *atopos*
accept: *prosiesthai*
accessible, readily: *prokheiros*
accomplished, be: *epiteleisthai*
accordance with nature: *to kata phusin*
account: *logos*
account for: *aitiologein*
accretion: *proskrisis*
accrue: *prosginesthai*
accurately: *akribôs*
accustom: *sunethizein*
acquainted with, be: *ginôskein*
acquire: *lambanein*
acquisition: *lêpsis*, *proslêpsis*
active: *energêtikos*
  be active: *energein*
activity: *energeia*
actual: *kat' energeian*
  actually, in actuality: *energeiai*, *kat' energeian*
acumen: *ankhinoia*
add: *epagein*, *epipherein*, *prostithenai*
  be added: *proseinai*, *proskeisthai*
addition: *prosthêkê*
additional specification: *prosdiorismos*
adduce: *epipherein*
adequate: *hikanos*
adjacent, be: *pelazein*
admirably: *thaumastôs*
admit: *epidekhesthai*, *sunkhôrein*
adult: *teleios*
adventitious: *epeisodiôdês*
affected, be: *paskhein*
  easily affected: *eupathês*
  subject to being affected: *pathêtikos*
affection: *pathos*
affective: *pathêtikos*
affinity: *sungeneia*
affirmative: *kataphatikos*
aforementioned: *proeirêmenos*
a fortiori argument, on the basis of an: *ek tou mallon*
aggregate: *sunkrinein*
aggregating: *sunkrisis*
aggregative: *sunkritikos*
agree: *homologein*
aim (n.): *skopos*
air: *aêr*
all, absolutely: *hapas*
alleged: *rhêtheis*
alter (trans.): *alloioun*
  altered, be: *alloiousthai*
alteration: *alloiôsis*
alternately: *enallax*
amber: *êlektron*
ambiguous, be: *kat' amphoterôn legesthai*
analogy: *analogia*
ancient: *arkhaios*
animal: *zôion*
animate: *empsukhos*
annoy: *enokhlein*
another: *allos*
  of another kind: *heteros*
apart from: *khôris*
appear, be apparent: *phainesthai*
appearance: *phantasia*, *to phainesthai*
append: *paragraphein*, *prosgraphein*
appetite: *epithumia*
apprehend: *antilambanesthai*
apprehension: *antilêpsis*
appropriate: *oikeios*
  be appropriate: *prosêkein*
argue syllogistically: *sullogizesthai*
argument: *epikheirêma*, *logos*
arise: *enginesthai*
arrive at: *ginesthai* (with *en*)
articulate: *diarthroun*
  detailed articulation: *diarthrôsis*
ascend: *anabainein*
ask: *phanai*

asleep, be: *katheudein*
aspirate: *dasunein*
assert: *phanai*
assign: *tattein*, *tithenai*
assimilate: *exomoioun*
assimilative: *lêptikos*
associated with, be: *suneinai*
assume: *lambanein*
  assume in addition: *proslambanein*
  assume in advance: *prolambanein*
assuredly: *pantôs*
astray, go: *periplanasthai*
at the same time: *hama*
atemporally: *akhronôs*
atom: *atomos*
attack: *ephistanein*
authoritative: *kurios*
awake, be: *egeirein*
axiom: *axiôma*

bad: *kakos*, *mokhtheros*
bad temperament: *duskrasia*
badly: *kakôs*
be: *einai*, *huparkhein*
  be present: *eneinai*, *pareinai*
bear (vb.): *ekhein*
beauty: *kallos*
become: *ginesthai*
bed: *klinê*
befit: *prosêkein*
beg the question: *to en arkhêi aitein*
begin: *arkhesthai*
beginning: *arkhê*
being: *on*, *to einai*
believe: *nomizein*
belong: *huparkhein*
beloved: *erastos*
better: *ameinon*, *kallion*
between, in between: *metaxu*, *en mesôi*
bind in: *endein*
bitterness: *pikrotês*
black: *melas*
  turn black: *melainesthai*
  turning black: *melansis*
blessed: *makaros*
blow, deliver a: *plêttein*
boast, make a: *kompazein*
bodily: *sômatikos*
  in a bodily way: *sômatikôs*
body: *sôma*
bone: *ostoun*
book: *biblion*
borne, be: *pheresthai*
boxing, capable of: *puktikos*
brazen: *khalkous*
bring: *agein*
  bring into collision: *sunkrouein*
  bring to mind: *hupomimnêskein*
  bring together: *sunagein*
  which brings together: *sunagôgos*
bronze: *khalkos*
bury: *katakhônnunai*

call: *kalein*
calm, calming: *galênê*
can: *dunasthai*
candle: *puramis*
capacity: *dunamis*
carry: *okhein*
  carry along: *pherein*
  (the) carrier: *to okhoun*
  carrying: *okhêsis*
case, be the: *huparkhein*
category: *katêgoria*
cause (n.): *aitia*, *aition*
cause alteration: *alloioun*
  capable of causing alteration: *alloiôtikos*
cause change: *metaballein*
cause diminution: *phthinein*
cause increase: *auxein*
cause locomotion: *pherein*
cease: *pauesthai*
chaff: *akhuron*
chance: *tukhê*
  a (any) chance (thing): *to tukhon*
change (n.): *metabolê*
change (vb.): *metaballein*
  easily changeable: *eumetablêtos*
  easily changed: *eumetabolos*
characterize: *kharakterizein*
child: *pais*
  little child: *paidion*
choose: *hairesthai*
circle: *kuklos*
  the circle of: *hoi peri*
circuit wall: *peribolos*
circular locomotion: *kuklophoria*
cite: *paratithenai*
  citing instances: *parathesis*
city: *polis*
civic: *politikos*
claim (vb.): *legein*
clarify: *saphênizein*
classify under: *anagein hupo*
clear: *saphês*
coexist: *sunuparkhein*
cognition: *gnôsis*
cognizable: *gnôstos*
coincide: *epharmottein*
cold: *psukhros*
coldness: *psukhrotês*
collision, bring into: *sunkrouein*

colour (n.): *khrôma*
colour (vb.): *khrôizein*
combine: *suntithenai*
combining: *sunthesis*
come: *erkhesthai*
come on: *epienai*
come to a halt: *histasthai*
come to be, come into being: *ginesthai*
come to be present: *paraginesthai*
come to exist: *sunistasthai*
coming to be: *genesis*
commentary: *hupomnêma*
commentator: *exêgêtês*
common, in common: *koinos*
 common feature: *koinotês*
comparable: *sumblêtos*
compare: *paraballein*, *sumballein*, *sunkrinein*
 make a comparative judgment: *sunkrinein*
comparison: *parabolê*, *sunkrisis*
complete (vb.): *sumplêroun*
completely: *teleiôs*, *teleôs*
complexion: *khrôma*
composed, be: *sunkeisthai*
comprehend: *perilambanein*
compress: *sunkrinein*
 which tends to compress: *sumpilêtikos*
compressing: *sunkrisis*
concise: *suntomos*
concisely: *suntomôs*
conclude: *sumperainesthai*
conclusion: *sumperasma*, *sunagôgê*
 draw the conclusion: *epipherein*, *sumperainesthai*
condensed, be: *puknousthai*
condition: *diathesis*
 good condition: *euexia*
 be in good condition: *euthêmoneuein*
 natural condition: *to kata phusin*
confirm: *pistousthai*
conjecture: *huponoia*
consequent (upon), be: *hepesthai*
consider: *skopein*, *theôrein*
 one must consider: *skepteon*
consideration: *katanoêsis*
consonant with, be: *sunaidein*
constituted, be naturally: *phuein* (perf.)
constituting: *sustasis*
construct: *sunistanai*
contact: *haphê*
contemplate: *theôrein*
content, be: *arkein* (pass.)
continuity: *sunekheia*
continuous: *sunekhês*
 make continuous: *sunekhizein*
contraposition: *antistrophê sun antithesei*
contrary way, in the: *enantiôs*
contribute: *sunteleioun*
conversely: *anapalin*, *antistrophôs*
conversion: *antistrophê*
 be convertible: *antistrephein*
cooled, be: *psukhesthai*
cooling: *psuxis*
coping: *thrinkos*
corollary: *porisma*
corporeal: *sômatikos*
 in a corporeal way: *sômatikôs*
corpse: *nekros*
correct: *orthos*
correct (vb.): *diorthoun*
correlate: *paraballein*
correlation: *parabolê*
corresponding respects, in: *katallêlôs*
corrupted, become: *kakunesthai*
courage: *andria*
cowardice: *deilia*
craft: *tekhnê*
 pertaining to crafts: *tekhnikos*
creation: *dêmiourgia*
criterion: *kritêrion*
criticize: *aitiasthai*
cured, be: *iasthai*
curve: *periphereia*
curved, curved line: *peripherês*
custom: *ethos*
cut: *temnein*

damningly: *entreptikôs*
darker, become: *melainesthai*
dead: *nekros*
deaden: *nekroun*
deceive: *diapseudein*
deception: *apatê*
declare: *eipein*
decrease (intrans.), be decreased: *meiousthai*
 cause decrease: *meioun*
decrease (n.): *meiôsis*
deduce syllogistically: *sullogizesthai*
deem proper, deem worthwhile: *axioun*
deeply: *dia bathous*
defective: *ellipês*
defend: *paristanai*
 defend oneself: *apologeisthai*
defence: *sunêgoria*
deficiency: *elleipsis*
define: *diorizein*, *horizesthai*
definite: *hôrismenos*
definitely: *hôrismenôs*
definition: *horismos*, *logos*
deliver a blow: *plêttein*

demonstrate: *apodeiknunai*
demonstration: *apodeixis*
deny: *anairein, apophaskein*
depart from: *apoleipein*
departure: *ekstasis*
depend: *artasthai*
deprive: *sterein*
derive (a word): *etumologein*
descend: *katabainein*
desiderative: *orektikos*
designate: *prosagoreuein*
desire: *orexis*
desired: *orekton*
destroy: *phtheirein*
destructive: *anairetikos*
detach: *apospan, diairein*
detailed articulation: *diarthrôsis*
determined: *hôrismenos*
  be determined: *horizesthai*
diagonal: *diametros*
differ: *diapherein*
difference: *diaphora*
  make a difference: *diapherein*
different: *diaphoros, heteros*
  of different natures: *heterophuês*
  of different species: *anomoeidês*
differentia: *diaphora*
dilate: *diakrinein*
dilating: *diakrisis*
diminish (intrans.), cause diminution: *phthinein*
directly (sc. proving): *deiktikôs*
discover: *heuriskein*
discussion: *logos*
disdain: *kataphronein*
dismiss: *aphienai*
  to be dismissed: *apoblêtos*
display: *apodidonai, ekhein* (with *diaphora*)
dispose: *diatithenai*
disproportion: *asummetria*
disproportionate: *asummetros*
dispute: *antilegein*
dissimilar: *anomoios*
dissimilarity: *anomoiotês*
distance: *diastêma*
distinct: *diôrismenos*
distinction: *diakrisis, diorismos*
distinctive: *idios*
distinguish: *diairein, diakrinein, diorizein*
distinguishing: *diakritikos*
disturb: *tarattein*
disturbance: *tarakhê*
diverge from: *apokhôrein*
divide: *diairein, dialambanein*
  divide as well: *sundiairein*
divisible: *diairetos*
division: *diairesis*
  by division: *ek diaireseôs*
do: *dran, poiein, prattein*
  do rightly: *katorthoun*
doctrine of nature: *phusiologia*
dog: *kuôn*
dominate: *kratein*
double: *diplasios*
doubleness: *to diplasion*
downhill: *katantês*
draw: *agein*
  draw the conclusion: *epipherein, sumperainesthai*
  draw (water): *arutein*
  draw away: *apagein*
  draw toward: *prosagein*
drawing in: *holkê*
dried, be: *xêrainesthai*
drink: *poma*
drunkenness: *methê*
  be drunk: *metheuein*
dry: *xêros*
dryness: *xêrotês*
due measure: *metron*
  lack of due measure: *ametria*
due proportion: *summetria*
  due proportionality: *to summetron*
  duly proportioned: *summetros*

each: *hekastos*
earlier (adv.): *proteron*
earth: *gê*
education process: *paideusis*
effectively: *energês*
efficient (cause), the: *to poiêtikon*
  in the manner of an efficient cause: *poiêtikôs*
effluence: *aporrhoia*
element: *stoikheion*
eliminate: *anairein*
eliminative: *ekkritikos*
emotion: *thumos*
employ: *paralambanein*
encompass: *periekhein*
end: *peras, telos*
endure: *hupomenein*
entailment: *akolouthia*
entangled, be: *emplekesthai*
entrust: *epitrepein*
enumerate: *aparithmein*
equal: *isos*
  equal in speed: *isotakhês*
  equality: *isotês, to ison*
  equally: *homoiôs*
eradicate: *ekkoptein*
error: *planê*

escape: *apopheugein*
escape notice: *lanthanein*
essence: *ousia, to einai, to ti ên einai*
  invest something with essence: *ousioun*
  the essence of X: *to X einai*
essential: *ousiôdês*
establish: *kataskeuazein, sunistanai*
everlasting: *aïdios*
every, absolutely: *hapas*
everywhere: *pantakhou*
evident: *enargês, phaneros*
evidently: *enargôs, phanerôs*
exact: *akribês*
exactly: *akribôs*
examination: *exetasis*
examine: *episkopein*
  examine together: *sunexetazein*
example: *paradeigma*
exceed: *huperekhein*
excellent: *spoudaios*
excess: *huperbolê*
exchange (vb.): *allassein*
excretion: *apokrisis*
exhaling: *ekpnoê*
exhausted, be: *ekluesthai*
exist: *einai, huparkhein, sunistasthai*
  come to exist: *sunistasthai*
existence: *huparxis, hupostasis*
  capable of bringing (something) into existence: *hupostatikos*
expect: *elpizein*
expectation: *elpis*
experience: *empeiria*
explain: *exêgeisthai*
explanation: *exêgêsis*
extend: *ekteinein*
extremity: *akron*
extrinsic: *epeisaktos*
eye: *ophthalmos*

faint: *amudros*
fall: *katapiptein*
  fall under: *piptein* (with *eis*)
fallacy: *paralogismos*
far removed, be: *apekhein*
fast: *takhus*
fastness: *to takhu*
father: *patêr*
fault with, find: *enkalein*
feature, common: *koinotês*
  incidental feature: *sumbebêkos*
fewer: *elattôn*
fiery: *purios*
figure: *morphê*
figure (of inference): *skhêma*
final: *teleutaios*
final (cause), the: *to telikon*
find: *heuriskein*
  find fault with: *enkalein*
  be found: *keisthai*
finite: *peperasmenos*
fire: *pur*
first: *prôtos*
first (adv.): *prôton, prôtôs*
first mover: *prôton kinoun*
fit (vb.): *harmottein*
five-talent: *pentetalantos*
fixed: *araros*
flat: *huptios*
flavour: *khumos*
flawless: *aptaistos*
flesh: *sarx*
flow (vb.): *rhein*
  flow away: *aporrhein*
fly: *hiptasthai*
flying: *ptêsis*
follow, follow from: *akolouthein, hepesthai*
  follow after: *epakolouthein*
  follow closely: *parakolouthein*
  follow mutually from: *antakolouthein*
folly: *aphrosunê*
food: *trophê*
foot: *pous*
force: *bia, dunamis*
  be forced: *biazesthai*
  forced: *biaios*
  forcibly overpower: *biazesthai*
form: *eidos*
  constitute (something's) form, *eidopoiein*
  constitutive of form: *eidopoios*
formulate (an argument): *sunagein*
found, be: *keisthai*
fraction: *pollostêmorion*
furnish: *parekhesthai*
further on: *proelthôn* (from *proerkhesthai*)
fuss with: *polupragmonein*
future, in the: *loipon*

gain (vb.): *apolambanein*
gaining: *apolêpsis*
gap, leave a: *dialeipein*
gather: *athroizesthai*
general: *katholikos, koinos*
  in general: *holôs*
  in a general sense: *koinôs*
generalize: *koinopoiein*
generally: *koinôs*
  speaking generally: *holôs*
generate: *gennan*
generative: *gennêtikos*
generic: *genikos*
genuinely: *ontôs*

genus: *genos*
  of the same genus: *homogenês*
germane: *oikeios*
give: *apodidonai, paradidonai*
  give as the cause: *aitiasthai*
  give by way of contrast: *antapodidonai*
  give up on: *apoginôskein*
glorious: *kudimos*
go: *ienai*
  go astray: *periplanasthai*
  go on, over: *epienai*
  go out of being: *apoginesthai*
  go through: *diienai*
good: *agathos*
good condition: *euexia*
grammarian: *grammatikos*
grasp: *lambanein*
great: *megas, polus*
  be greater: *pleonazesthai*
grey: *phaios*
grind grain: *alêthein*
ground: *edaphos*
growth: *auxê*
guide aright: *apeuthunein*

habit: *ethos*
  make a habit of: *ethein*
habituation: *sunethismos*
hair: *thrix*
half-hour: *hêmiôrion*
halt: *stasis*
  come to a halt: *histasthai*
hand: *kheir*
happen: *sumbainein, tunkhanein*
happy, be: *eudaimonein*
hard: *sklêros*
harmony, be in: *sumphônein*
  in harmony: *sumphônos*
haul (a ship): *neôlkein*
have: *ekhein*
health: *hugeia*
healthy: *hugiês*
  be healthy: *hugiainein*
  become healthy: *hugiazesthai*
  getting healthy: *hugiansis*
hear: *akouein, kluein*
hearing: *akoê*
heart: *kardia*
heat red-hot: *puraktoun*
heated, be: *thermainesthai*
heating: *thermansis*
heavy: *barus*
help (n.): *boêtheia*
help (vb.): *boêthein*
hence: *toigaroun, dia touto*
high note: *nêtê*
highest: *anôtatô*
hinder, be a hindrance: *empodizein*
hit upon: *tunkhanein*
hold down: *katekhein*
  hold in check: *epekhein*
  lay hold of: *drattesthai*
  which hold together: *sunektikos*
homoeomerous: *homoiomerês*
homonymous: *homônumos, homônumon*
  homonymous meaning: *homônumia*
homonymously: *homônumôs*
horse: *hippos*
hot: *thermos*
hotness: *thermotês*
hour: *hôra*
  half-hour: *hêmiôrion*
house: *oikia, oikos*
human: *anthrôpeios, anthrôpinos*
human being: *anthrôpos*
hundred times, a: *hekatontaplasios*
hundredth: *hekatostos*
hypothesis: *hupothesis*
hypothesize: *hupotithesthai*
  be hypothesized: *hupotithesthai, hupokeisthai*
hypothetical: *hupothetikos*

ignorance: *anepistêmosunê*
ignore: *katanôtizesthai*
illumine: *phôtizein*
illustration, setting out an: *ekthesis*
illustrative: *ekthetikos*
image: *eikôn*
imaginative: *phantastikos*
imitate: *mimousthai*
immediately: *amesôs, autothen, euthus*
  immediately after: *ephexês*
immortal: *athanatos*
impart motion: *kinein*
impassivity: *apatheia*
imperfect: *atelês*
important: *kurios*
impossible: *adunatos*
impulse: *hormê*
in actuality: *energeiai, kat' energeian*
in between: *metaxu, en mesôi*
in common: *koinos, koinôs*
in general: *holôs*
in harmony: *sumphônos*
  be in harmony: *sumphônein*
in its own right: *kath' hauto, kath' heauto*
inanimate: *apsukhos*
incidental: *sumbebêkos*
  incidental, incidentally: *kata sumbebêkos*
  incidental feature: *sumbebêkos*
inclination: *rhopê*

incline: *apoklinein*
include: *suntattein*
incommensurable: *asummetros*
incorporeal: *asômatos*
increase (intrans.), be increased: *auxesthai*
  increase (trans.), cause increase: *auxein*
increase (n.): *auxêsis*
indefinite: *aoristos*
  in an indefinite way: *adioristôs*
indicate: *dêloun*
individual (n.): *atomos*, *to kath' hekaston*
  individual type: *idiotropia*
  individually: *idiai*
indivisible: *adiairetos*, *atomos*
induction: *epagôgê*
  based on induction: *ek tês epagôgês*
  by induction: *apo tês epagôgês*
  through induction: *dia tês epagôgês*
inductively: *epaktikôs*
inequality: *anisotês*
inexperienced: *apeiros*
infer: *epagein*, *sunagein*
inference: *sunagôgê*
  make the inference: *sunagein*
inferior: *kheirôn*
infinite: *apeiros*
  infinitely many, infinite in multitude: *apeiroi tôi plêthei*
  to infinity: *ep' apeiron*
inflow, subject to: *epirrhutos*
inhaling: *eispnoê*
inhere: *enuparkhein*
inopportunely: *akairôs*
inquire, inquire about: *zêtein*
  inquire further into: *epizêtein*
inquiry: *zêtêsis*
insensate: *anaisthêtos*
inserted, be: *parempiptein*
insofar as: *katho*
inspection: *ephodos*
instrument: *organon*
intellect: *nous*
intellectual: *noêtikos*
intelligent: *phronimos*
  lacking intelligence: *anous*
  lack of intelligence: *anoia*
intemperance: *akolasia*
interject: *paremballein*
intermediary (n.): *meson*
intermediate: *mesos*
interrupt: *diakoptein*
interval: *diastasis*
introduce: *epagein*
invariably: *pantôs*
inverse: *antikeimenon*
investigate: *zêtein*
involve: *ekhein*
iron: *sidêros*
irrational: *alogos*
itself, by: *kath' heauto*

join: *sunaptein*
judge: *krinein*, *tekmairesthai*
  judge together: *sunkrinein*
judgment: *krisis*
just stated: *proeirêmenos*
justice: *dikaiosunê*
justly: *dikaiôs*

keep: *phulattein*
kharistion (an instrument for weighing): *kharistiôn*
kind: *genos*
  of another kind: *heteros*
  of two kinds: *dittos*
kindled, be: *exaptesthai*
know: *eidenai*, *epistasthai*, *ginôskein*, *gnôrizein*
  one ought to know: *isteon*
knowable: *epistêtos*
knowing (adj.), knower: *epistêmôn*
knowledge: *epistêmê*
known, well-known: *gnôrimos*

label: *epigraphein*
lack of intelligence: *anoia*
lacking intelligence: *anous*
lame: *khôlos*
larger: *meizôn*
last: *eskhatos*
later (adv.): *husteron*
lay hold of: *drattesthai*
laying round: *peribolê*
layperson: *idiôtês*
lead (n.), white: *psimuthion*
lead (vb.): *hêgeisthai*
learn: *manthanein*
learning: *mathêsis*
least: *elakhistos*
leave a gap: *dialeipein*
leave off: *lêgein*
leave out: *paraleipein*
length: *mêkos*
less: *elattôn*, *hêttôn*
  be less: *elattousthai*
level, on the same: *sustoikhos*
lie (vb.): *keisthai*
  lie on: *epikeisthai*
life: *zôê*
  stage of life: *hêlikia*
  way to live: *bios*
light (adj.): *kouphos*

light (n.): *phôs*
like, be: *proseoikein*
limit: *horos*
line: *grammê*
  curved line: *peripherês*
  straight line: *eutheia*
liquid: *hugros*
live, way to: *bios*
load: *phortion*
locomotion: *phora*
  move locally, be moved locally: *pheresthai*
  cause locomotion: *pherein*
  circular locomotion: *kuklophoria*
  rectilinear locomotion: *euthuphoria*
log: *xulon*
logical: *logikos*
look: *blepein*
lose: *apoballein*
loss: *apobolê*
  easily lost: *euapoblêtos*

made, be: *ginesthai*
magnitude: *megethos*
main point: *kephalaion*
maintain: *phulattein*
major (premise): *meizôn*
make: *poiein*, *apodidonai*
  make a (the) sound: *psophein*
  make a distinction: *diorizein*
  make continuous: *sunekhizein*
  make ready: *prokheirizesthai*
  make sense: *ekhein logon*
  make the inference: *sunagein*
  make use of: *apokhrasthai*, *sunkrasthai*
man: *anêr*
manifest: *prophanês*
manifestation: *ekphansis*
manner: *tropos*
manuscript: *antigraphon*
many: *polus*
  in many cases: *pollakhou*
  in many places: *pollakhou*
  many times (as great as): *pollaplasios*
  in many ways: *pollakhêi*
master: *kathêgemôn*
matter: *hulê*
mean: *sêmainein*, *legein*
measure (n.): *metron*
measure (vb.): *metroun*
medimnus: *medimnos*
medium: *meson*
meet with: *hupantan*
memory: *mnêmê*
mental: *dianoêtikos*
mention: *hupomimnêskein*, *mnêmoneuein*
method: *methodos*
  methodically: *hodôi*
middle: *mesos*
  in the middle of: *ana meson*
millet seed: *kenkhros*
millstone: *mulos*
  (in a handmill): *mulê*
mind: *dianoia*, *ennoia*
minor (premise): *elattôn*
missing, be: *elleipein*
moderate (vb.): *metroun*
  moderate (adj.): *memetrêmenos*
  moderately: *memetrêmenôs*
moistened, be: *hugrainesthai*
molten, have become: *kheisthai* (pf.)
moral: *êthikos*
mortal: *thnêtos*
motion: *kinêsis*
  impart motion: *kinein*
move (intrans.), be moved: *kineisthai*
  move (something): *kinein*
  move along with: *sunkineisthai*
  move locally: *pheresthai*
  capable of moving (something): *kinêtikos*
movement: *kinêma*
mover: *kinoun*
  first mover: *prôton kinoun*
  proximate mover: *prosekhôs kinoun*
much: *polus*
muchness: *to polu*
multitude: *plêthos*
music: *mousikê*
must: *dein*, *khrê*
mutable, easily: *eutreptos*

nail: *onux*
name (vb.): *onomazein*
nameless: *anônumos*
natural: *phusikos*, *kata phusin*
  naturally: *phusikôs*, *kata phusin*
  be naturally, be naturally constituted: *phuein* (perf.)
  be naturally well-suited: *euphuôs ekhein*
nature: *phusis*
  in accordance with nature: *kata phusin*; accordance with nature, natural condition, natural: *to kata phusin*
  contrary to nature: *para phusin*
  doctrine of nature: *phusiologia*
  having to do with nature: *phusikos*
  of different natures: *heterophuês*
  of like nature: *homophuês*
necessary: *anankaios*, *anankê*
  it is necessary: *anankaios*, *anankê*, *khrê*, *khrênai*
  necessitate: *anankazein*
necessity: *anankê*

need: *deisthai*
needful: *deon*
neglectful of: *katamelês*
next: *ephexês, hexês, loipon, ekhomenon*
nicely: *kalôs*
noncomparability: *to asumblêton*
noncomparable: *asumblêtos*
nonhomoeomerous: *anomoiomerês*
nonknower: *anepistêmôn*
note (n.), high: *nêtê*
  next-to-the-highest note: *paranêtê*
note (vb.): *ephistanein*
novel way, in a: *kainoprepôs*
nowhere: *oudamou*
number: *arithmos*
  in number: *arithmôi*
  numerically: *kat' arithmon*
nurture, nutrition: *trophê*

objection: *enstasis*
  object, raise an objection: *enistasthai*
obliged, be: *opheilein*
observe: *theôrein*
obvious: *dêlos*
  obviously: *prodêlôs*
  obvious in advance, quite obvious: *prodêlos*
occur: *ginesthai*
odds, be at: *diaphônein*
of another kind: *heteros*
offer: *paradidonai*
omit: *parienai*
on the right: *dexios*
only: *monos, monôs*
opposed, opposite: *antikeimenon*
order: *taxis*
organ: *organon*
  organ of taste: *geusis*
  organ of smell: *osphrêsis*
original: *ex arkhês, pothen arkhomenê*
  originally: *ex arkhês*
original (n.): *prôtotupon*
otherness: *heterotês*
otherwise (adj.): *alloios*
otiose: *perittos*
ought: *dein*
outflow, subject to: *aporrhutos*
outside, from: *exôthen*
overpower: *katiskhuein*
own (adj.): *idios*
  own, own proper: *oikeios*
  in its own right: *kath' hauto, kath' heauto*

pain: *lupê*
  cause pain: *lupein*
  be pained: *lupesthai*
painful: *lupêros*
paler, become: *leukainesthai*
paraphrase: *paraphrazein*
paronymously: *parônumôs*
part: *meros, morion*
partake (of): *metekhein*
particular: *merikos*
  the particular: *to en merei, to kata meros, to merikon*
parting the warp: *kerkisis*
partition: *merizein*
partless: *amerês*
passage: *metabasis*
passage (of a text): *lexis, rhêsis*
pass by: *parerkhesthai*
pass over: *paratrekhein*
pass through: *diaporeuesthai*
pass to, pass on to: *metabainein*
passenger: *epibatês*
pencil: *grapheion*
penetrate: *khôrein* (with *dia bathous*)
perceive: *aisthanesthai*
perception: *aisthêsis*
  capable of perception: *aisthêtikos*
perchance: *ei tukhoi*
perfect: *teleios, holoklêros*
perfect (vb.): *teleioun*
perfecting: *teleiôsis*
perfection: *teleiotês, to teleion*
perform: *apotelein, poieisthai*
perhaps: *isôs, takha*
perish: *phtheirein* (mid.), *apollusthai*
perishing: *phthora*
person: *anthrôpos*
pertain: *huparkhein*
philosopher: *philosophos*
  in philosophical fasion: *philosophôs*
physicist: *phusikos*
place: *topos*
plainly, be: *phainesthai* (with ptc.)
plant: *phuton*
pleasant: *hêdus*
please: *hêdein*
pleasure: *hêdonê*
pneuma: *pneuma*
point: *sêmeion*
  main point: *kephalaion*
  point proposed: *to prokeimenon*
point out: *deiknunai, paradeiknunai*
portion: *morion*
pose (an argument, a question): *eresthai, erôtan*
  pose (a puzzle): *aporein*
posit: *tithenai*
  be posited: *keisthai*
possess: *ekhein*

the possessor: *to ekhon*
possible: *dunatos, endekhomenos*
be possible: *dunasthai, endekhesthai, einai*
as a possibility: *hôs endekhomenos*
potentiality: *dunamis*
potential: *to dunamei, to kata dunamin*
potentially: *dunamei*
power: *dunamis*
precede: *prolambanein*
preconception: *prolêpsis*
predicate (vb.): *katêgorein*
predication: *katêgoria*
premise: *protasis*
preparation: *paraskeuê*
prepare: *paraskeuazein*
prepared (for): *hetoimos*
present, be: *eneinai, pareinai*
come to be present: *paraginesthai*
present (vb.): *proagein*
preserve: *sôizein, diasôizein*
prevail, prevail over: *epikratein*
prevent: *kôluein*
primarily: *prôtôs*
primary: *prôtos*
principle: *arkhê*
problem: *problêma*
proceed: *ienai, meterkhesthai, proerkhesthai, proïenai, prokhôrein*
procession: *proödos*
produce: *apotelein, poiein*
produce in: *empoiein*
be produced: *ginesthai*
productive: *poiêtikos*
prolong: *mêkunein*
proof: *deixis*
in proof of: *deiktikos*
proper: *oikeios*
deem it proper: *axioun*
properties: *ta huparkhonta*
proportion: *analogia*
proportional, in proportion: *analogos*
proportionally: *analogôs*
due proportion: *summetria*
duly proportioned: *summetros*
due proportionality: *to summetron*
propose: *protithenai, prokeisthai*
be proposed: *prokeisthai*
point proposed: *to prokeimenon*
propound: *apodidonai, erôtan*
prove: *deiknunai*
prove before: *prodeiknunai*
proximate (i.e. immediately next): *prosekhês*
proximately: *prosekhôs*
proximate mover: *prosekhôs kinoun*
psychic: *psukhikos*
pull: *helkein*
pull along after: *sunephelkein*
pull along: *ephelkesthai*
puller, the: *to helkon*
pulling: *helxis, holkê*
purely: *katharôs*
purport: *ennoia*
purpose: *hou kharin*
push: *ôthein*
push along: *epôthein*
push apart: *diôthein*
push away: *apôthein*
push together: *sunôthein*
pushing: *ôsis*
pushing along: *epôsis*
pushing apart: *diôsis*
pushing away: *apôsis*
pushing together: *sunôsis*
put: *tithenai, diatattesthai*
put before: *protithenai*
put forward: *proballein, protithenai*
put succinctly: *sunairein*
putting on: *epithesis*
puzzle: *aporia*
pose a puzzle: *aporein*
be thoroughly puzzled: *diaporein*
puzzling: *aporos*

quality: *poiotês, poion*
qualified thing: *poion*
quantity: *posotês, poson*
quickly: *takheôs*
quiet down: *hêsukhazesthai*

random, at: *hôs etukhe*
rarefied, be: *manousthai*
rather: *mallon*
ratio: *analogia, logos*
rational: *logikos*
pertaining to rational disciplines: *logikos*
reacquire: *analambanein*
read: *entunkhanein*
readily accessible: *prokheiros*
ready, make: *prokheirizesthai*
ready to hand: *prokheiros*
really: *ontôs*
in reality: *tôi onti*
reason (i.e. the faculty): *logos*
reasonably: *eikotôs*
reason why: *aitios*
with good reason: *eulogôs*
receive: *dekhesthai, tunkhanein*
receive in succession: *diadekhesthai*
capable of receiving: *dektikos*

recipient: *to dektikon, to dekhomenon, to dedegmenon*
recent: *neos*
reciprocal replacement: *antiperistasis*
reckon in addition: *proslogizesthai*
recollection: *anamnêsis*
rectilinear locomotion: *euthuphoria*
reduce: *apagein*
 reduce to: *anagein eis*
 reduction to impossibility: *apagôgê eis adunaton*
redundant, be: *parelkein*
refer (someone) to: *anapempein*
refer to paronymously: *parônumiazein*
refute: *dielenkhein, elenkhein, luein*
reject: *elenkhein*
relationship: *skhesis*
relative: *pros ti*
 be related to: *ekhein pros*
 relativity: *to pros ti*
remain, remain stationary: *menein*
remaining: *loipos*
 be remaining: *hupoleipesthai*
remark: *ephistanein*
remember: *mimnêskesthai*
remind: *hupomimnêskein*
removal: *exairesis*
render: *apotelein*
reply (vb.): *anthupopherein*
report: *apangellein*
require as a matter of necessity that: *anankazein*
resemble: *epeoikein*
reshaping: *metaskhêmatisis*
resistance, involving: *antereistikos*
resolve: *dialuein*
responsible: *aitios*
rest: *êremia*
 be at rest: *êremein*
 come to rest: *êremizesthai*
 coming to rest: *êremêsis, êremisis*
rest, the: *to loipon*
restored, be: *apokathistasthai*
result: *sumbainein*
return: *epanienai*
revolution: *periodos*
rid of, get: *ekballein*
rider: *anabatês*
right: *euthus*
right, in its own: *kath' hauto, kath' heauto*
right, on the: *dexios*
rightly: *kalôs*
river: *potamos*
rock: *petra*
rod: *rhabdos*
role: *logos*
roof tile: *keramos*
 roof with tiles: *keramoun*
rope: *skhoinos*
rotation: *periphora*
rule: *kanôn*
run with a fair wind: *ouriodromein*
running, capable of: *dromikos*

safe: *asphalês*
sail: *plein*
sake of, for the: *heneka, heneken, kharin*
 that for the sake of which: *to hou heneka, to hou heneken*
same: *autos*
sameness: *tautotês*
say: *eipein, legein, phanai*
scholarly: *philologos*
scientifically: *epistêmonikôs*
see: *horan*
seeing: *anablepsis*
seek: *zêtein*
seem: *dokein, eoikein*
segregate: *diakrinein*
 segregating: *diakrisis*
 segregative: *diakritikos*
sense: *aisthêsis*
 capable of sensation: *aisthêtikos*
sense, make: *ekhein logon*
sense organ: *aisthêtêrion, aisthêsis*
sensible, sensible thing: *aisthêtos*
sensory: *aisthêtikos*
sentence: *logos*
separable: *khôristos*
separate: *khôrizein, apokhôrizein*
 separated: *kekhôrismenos*
separately: *khôristôs*
separative: *khôristikos*
set (something) out: *ektithesthai*
 be set forth: *ekkeisthai*
 set alongside: *paraballein*
 set free: *methienai*
 set in motion: *kinein*
 set in order: *diakosmein*
settle down: *kathistasthai*
 settling down: *katastasis*
several: *polus* (compar.)
shape: *skhêma*
 be shaped: *skhêmatizesthai*
shaping: *skhêmatisis*
sharp: *oxus*
sharpness: *to oxu*
shifted, be: *metakeisthai*
ship: *naus*
 haul (a ship): *neôlkein*
 ship-hauler: *neôlkos*
should not: *ouk ekhrên*

show: *deiknunai, paradeiknunai*
  be shown to be: *phainesthai* (with ptc.)
sick, be: *nosein*
sickness: *nosos*
side: *pleura*
  to the side: *epi to plagion*
  sideways: *plagios*
sight: *opsis*
sign: *sêmeion*
signify: *sêmainein*
similar: *homoios*
similarity: *homoiotês, to homoion*
similarly: *homoiôs*
  to a similar degree: *homoiôs*
simple: *haploïkos*
simply: *haplôs*
simultaneous, simultaneously: *hama*
sinew: *neuron*
sleep: *hupnos*
slow: *bradus*
slowly: *bradeôs*
slowness: *to bradu*
small: *brakhus*
smaller: *elattôn*
smell, organ of smell: *osphrêsis*
snow: *khiôn*
sober, be: *nêphein*
so-called: *kaloumenos*
solve: *luein*
Sophist: *sophistês*
soul: *psukhê*
sound: *phthongos, psophos*
  make a (the) sound: *psophein*
source: *arkhê*
speak, speak of: *eipein*
  speak before: *proerein*
  speak of, speak of as: *legein*
  speaking generally: *holôs*
species: *eidos*
  of different species: *anomoeidês*
  of the same species: *homoeidês*
specification, additional: *prosdiorismos*
specify: *diorizein*
speculation: *theôria*
  subject for speculation: *theôrêma*
speed, equal in: *isotakhês*
  of the same speed: *homotakhês*
  with equal speed: *isotakhôs*
spitting: *ptusis*
spring: *pêgê*
spur: *nuxis*
squaring (of the circle): *tetragônismos*
stade: *stadion*
  half a stade: *hêmistadion*
  of a stade: *stadiaios*
stage of life: *hêlikia*
stand: *bainein, histanai* (pf. act.)
standard: *horos*
starting here: *enteuthen*
state (n.): *hexis*
state (vb.): *eipein, legein*
  stated: *rhêtheis*
  just stated, stated previously: *proeirêmenos*
  statement: *rhêton, to eirêmenon*
  the preceding statements: *ta proeirêmena*
stationary, remain: *menein*
statue: *andrias*
stone: *lithos*
straight: *euthus*
  straight line: *eutheia*
straighten out: *kateuthunein*
stream: *rheuma*
strength: *iskhus*
  have the strength: *iskhuein*
strengthen: *kratunein*
stretching out: *ektasis*
  be stretched out alongside: *parateinesthai, sumparateinesthai*
strict sense, in the: *kuriôs*
strike: *prosballein*
strong: *iskhuros*
subject for speculation: *theôrêma*
subject to being affected: *pathêtikos*
subsequent: *ephexês*
substance: *ousia*
substantial: *ousiôdês*
substitute: *metalambanein*
substitution: *metalêpsis*
subsume under: *hupagein*
subtract: *aphairein*
subtraction: *aphairesis*
succession: *diadokhê*
succinctly: *sunêirêmenôs*
  put succinctly: *sunairein*
suffer: *paskhein*
suffice, be sufficient: *arkein*
suggest: *hupomimnêskein, eipein*
suitability: *epitêdeiotês*
suited: *epitêdeios*
  be naturally well-suited: *euphuôs ekhein*
sum: *sunamphoteron*
summarize: *epitrekhein*
superficial: *epipolaios*
superficially: *epipolaiôs*
superior: *kreittôn*
supervene: *epiginesthai, episumbainein*
supply: *endidonai*
suppose: *hêgeisthai, hupolambanein*
surface: *epiphaneia*
surroundings: *to periekhon*
suspect: *hupopteuein*

suspicion: *hupopsia*
  under suspicion: *hupoptos*
swan: *kuknos*
sweet: *glukus*
sweetness: *glukutês*, *to gluku*
syllogism: *sullogismos*
syllogistically deduce: *sullogizesthai*
synonymous: *sunônumos*
synonymously: *sunônumôs*
synonymy: *sunônumia*
systematic discussion: *tekhnologia*

take: *lambanein*
  take along with: *sumparalambanein*
  take part in: *sunepilambanein*
  take up: *prokheirizesthai*
  take up again: *analambanein*
talent: *talanton*
talk about: *legein* (with *peri*)
tamping the woof: *spathêsis*
tangible: *haptos*
tastable: *geustos*
taste: *geusis*
teach: *didaskein*
teaching: *didaskalia*
tell: *eipe* (imperative)
temperament, bad: *duskrasia*
temperance: *sôphrosunê*
ten-thousandth: *muriostos*
term: *onoma*
text: *graphê*, *lexis*
theorem: *theôrêma*
theory: *theôria*
therefore: *ara*
thing: *pragma*
thing imparting motion: *to kinoun*
think: *oiesthai*
  think of: *ennoein*
  think up: *epinoein*
thinking: *noêsis*
thought: *noêma*
throw: *rhiptein*
thrower, the: *to rhipsan*
throwing: *rhipsis*
tile, roof: *keramos*
time: *khronos*
  at the same time: *hama*
tincture: *parakhrôsis*
together: *hama*
top: *strombos*
totally: *pantelôs*
touch (n.): *haphê*
touch (vb.): *haptesthai*, *ephaptesthai*
  touching: *epaphê*, *haphê*
trace out: *anikhneuein*
train in advance: *progumnazein*
transfer: *metatithenai*
transmitted, be: *pheresthai*
transparent: *diaphanês*
travel: *ienai*
traverse: *dierkhesthai*
treatise: *pragmateia*
trial: *peira*
triangle: *trigônon*
true: *alêthês*
truth: *alêtheia*
try: *peirasthai*
turning: *dinêsis*
  turn (something): *dinein*
turning black: *melansis*
  turn black: *melainesthai*
turning white: *leukansis*
  turn white: *leukainesthai*
twist and turn: *strephein*
two kinds, of: *dittos*
type: *tropos*

ultimate: *eskhatos*
unable: *adunatos*
  be unable: *adunatein*
unaffected: *apathês*
unclear: *adêlos*
  be unclear: *asapheian ekhein*
  unclearness: *asapheia*
unconvincing: *apithanos*
undergo: *huphistasthai*, *paskhein*
  undergo an affection: *paskhein*
underlying, be: *hupokeisthai*
  the underlying thing: *to hupokeimenon*
understand: *phronein*, *akouein*
understanding: *phronêsis*
undifferentiated: *adiaphoros*
unequal: *anisos*
  unequal in speed: *anisotakhês*
unintelligent: *anoêtos*
unit: *monas*
united, be: *henousthai*
unity: *to hen*
universal: *katholou*, *holos*
universally: *katholou*
unmoved: *akinêtos*
untrained: *agumnastos*
unworthy: *anaxios*
uphill: *anantês*
upright: *orthos*
usage: *khreia*
use (vb.): *khrêsthai*
  make use of: *apokhrasthai*, *sunkhrasthai*
usefulness: *khreia*
using: *khrêsis*

vain, in: *matên*

valid (logical term): *hugiês*
vanish: *apoleipein*
variation: *parallagê*, *parallaxis*
variety: *diaphora*
vehemence: *sphodrotês*
vehement: *sphodros*
versions, in two: *dikhôs*
vice: *kakia*
  become corrupted: *kakunesthai*
view: *opsis*, *epibolê*
  be (someone's) view: *areskein*
vinegar: *oxos*
virtue: *aretê*
  become virtuous: *aretousthai*
visible: *horatos*
visual: *optikos*
voice (n.): *phônê*
voice (vb.): *phthengesthai*
volume: *onkos*

wake up: *egeirein*
walk: *badizein*
walking: *badisis*
wall, circuit: *peribolos*
want: *boulesthai*
warp, parting the: *kerkisis*
water: *hudôr*
wax: *kêros*
  waxen: *kêrinos*
way: *hodos*, *tropos*
  way to live: *bios*
weak: *atonos*, *malthakos*
weighing (adj.): *stathmistikos*
weight: *baros*
well-known: *gnôrimos*
well: *kalôs*
wet: *hugros*
wetness: *hugrotês*
where (as category label): *pou*
white: *leukos*
  white, the white: *to leukon*
  turn white: *leukainesthai*
  turning white: *leukansis*
  whiteness: *leukotês*
white lead: *psimuthion*
whole, as a whole: *holos*
wicked: *phaulos*
wind: *anemos*
wing: *pterux*
wish: *boulesthai*
withdraw: *apokhôrein*
within, from: *endothen*
wood: *xulon*
wooden: *xulinos*
woof, tamping the: *spathêsis*
word: *logos*
wording: *lexis*
worse: *kheirôn*
worth: *axios*
worthwhile: *axiologos*
  deem it worthwhile: *axioun*
write: *graphein*

young: *neos*

Greek-English Index

An asterisk after a citation indicates that the word occurs in the title, a lemma, or a quotation.

adêlos, unclear, 1056,4*; 1062,23
adiairetos, indivisible, 1040,32
adiaphoros, undifferentiated, 1044,7*; 1092,16; 1095,21*-1096,1 passim; 1097,22
adioristôs, in an indefinite way, 1045,12; 1047,11
adunatein, be unable, 1080,19; 1105,30
adunatos, impossible, 1039,14 et passim
unable, 1078,21*
(subst.) impossibility, 1042,29 et passim
apagôgê eis adunaton, reduction to impossibility, 1042,18
di'adunatou, through [reduction to] impossibility, 1082,20; 1083,7
aêr, air, 1050,3 et passim
agathos, good, 1044,6*.7.9.11
agein, draw, 1055,18*.23*
bring, 1057,23; 1059,31
agumnastos, untrained, 1077,19
aïdios, everlasting, 1077,7
aisthanesthai, perceive, 1059,24-1061,4 passim; 1072,28.29; 1078,27.29; 1081,5
aisthanomenos, (subst.) perceiver, 1058,1
aisthêsis, perception, 1058,30-1059,22 passim; 1072,8 bis; 1074,28; 1075,16.24; 1098,11.13 bis.14.16
sense, sense organ, 1058,26*.28* bis.30.31; 1059,5.10.12.18*.20.21.26 bis; 1060,2; 1079,27; 1080,5.28
aisthêtêrion, sense organ, 1060,9.15
aisthêtikos, capable of sensation, 1058,24.26.27*.29; 1059,3.27
capable of perception, 1080,3
sensory, 1060,24; 1072,5*-1075,11 passim; 1080,8; 1081,1.2; 1113,15.17
aisthêtos (adj. and neut. n.), sensible, sensible thing 1058,9-1080,28 passim; 1113,5.16
[***aitein***]
to en arkhêi aitein, beg the question, 1040,9; 1041,32
aitia, cause, 1048,12 et passim
aitiasthai, give as the cause, 1092,32; 1100,27.33; 1105,13
criticize, 1039,13.14
aitiologein, account for, 1050,8
aitios, responsible, 1041,6; 1092,20-1099,25 passim
reason why, 1082,29
because, 1097,7*
(neuter subst.) cause, 1048,13
akairôs, inopportunely, 1108,4
akhronôs, atemporally, 1076,6.7
akhuron, chaff, 1055,10
akinêtos, unmoved, 1040,5; 1048,14; 1079,15
akoê, organ of hearing, 1060,17
hearing, 1060,21
akolasia, intemperance, 1044,8.10
akolouthein, follow, follow from, 1041,27 et passim
akolouthia, entailment, 1039,23
akolouthos (adj.), follow from, 1104,32
akouein, hear, 1059,2.26
understand, 1047,26; 1108,11
akribês, exact, 1036,11; 1037,8
akribôs, exactly, 1037,2; 1042,8; accurately, 1078,28; 1080,5
akroasis, lecture course (in the title of the *Physics*), 1036,2*.3 (see note on 1036,3)
akron, extremity, 1048,19*
alêtheia (with ***kata***), in truth, 1067,1; 1088,19; 1097,11
alêthein, grind grain, 1053,26
alêthês, true, 1047,14 et passim
alêtheuesthai, be truly described by a predicate, 1070,28*
allassein, (vb.) exchange, 1062,29
alloios, otherwise (adj.), 1064,15; 1081,18
emended to ***allos***, 1068,22
alloiôsis, alteration, 1043,32 et passim
alloiôtikos, capable of causing alteration, 1111,14
alloioun, cause alteration, 1048,23*; 1056,21*; 1057,4*-1061,25 passim; 1111,6-1112,30 passim

alter (trans.), 1111,8
alloiousthai, be altered, 1056,19* et passim
undergo (with cognate accusative), 1059,19; 1111,19
allôs, in another way, 1055,3*; 1113,8
allôs ... allôs, in one sense ... in another, 1088,28-9
alogos, irrational, 1074,2.3; 1077,17
amerês, partless, 1040,32
amesôs, immediately, 1060,12
ametria, lack of due measure, 1070,5.16; 1073,16
amudros, faint, 1064,3
anabainein, ascend, 1084,6
anabatês, rider, 1053,15
anablepsis, seeing, 1076,7.12
anagein, (with ***eis***) reduce to, 1049,24; 1050,19.26; 1052,8*.14.16; 1053,5.7.28; 1056,13; 1065,8; 1074,8.18; 1112,26
(with ***hupo***) classify under, 1049,17.18; 1050,14; 1051,23*.26*; 1052,27; 1067,7; 1112,27
anairein, deny, 1042,25; 1051,20*; 1085,6; 1106,11; 1115,28
eliminate, 1083,21; 1090,2
anairetikos, destructive, 1039,19*
anaisthêtos, insensate, 1058,25; 1059,5; 1061,2*.3 bis.4
analambanein, take up again, 1052,3
reacquire, 1080,20 (in 1080,22 read as ***lambanein***)
analogia, analogy, 1096,31*; 1097,1*; 1114,25
proportion, 1103,2-1106,23 passim; 1109,13-1111,26 passim; 1115,11-1116,13 passim
ratio, 1111,4
analogos, proportional, 1109,18*
in proportion, 1108,10*
analogon (adv.), in proportion, 1105,26; 1110,27
analogôs, proportionally, 1106,15; 1107,15; 1110,14.17; with ***ekhein***, be proportional, 1110,23
anamnêsis, recollection, 1079,11
anankaios, necessary, 1072,2*; 1111,3
ekhei to anankaion, is necessary, 1104,23
anankazein, require as a matter of necessity that, 1109,29
necessitate, 1110,25
anankê, necessity (sometimes translated 'necessary' or 'necessarily'), 1037,11* et passim
anantês, uphill, 1084,2
anapalin, conversely, 1093,22*; 1101,5
anapempein, refer (someone) to, 1044,15
anaxios, unworthy, 1036,19
andria, courage, 1044,10
andrias, statue, 1062,26; 1063,10
anemos, wind, 1053,20
anepistêmosunê, ignorance, 1078,23
anepistêmôn, nonknower, 1079,7
anêr, man, 1102,20*; 1105,29 bis; 1107,18; 1109,31; 1115,30
anikhneuein, trace out, 1037,10
anisos, unequal, 1045,7 et passim
(term mentioned) 'unequal', 1102,13; 1114,29
anisotakhês, unequal in speed, 1084,4; 1102,4.11; 1115,6.7
anisotês, (term mentioned) 'inequality', 1101,28
ankhinoia, acumen, 1036,19; 1090,1
anoêtos, unintelligent, 1077,23*
anoia, lack of intelligence, 1077,6
anomoeidês, of different species, 1083,6; 1086,3; 1099,22
anomoiomerês, nonhomoeomerous, 1067,16
anomoios, dissimilar, 1054,19.21.23; 1101,2; 1102,17; 1114,33
(term mentioned) 'dissimilar', 1114,29
anomoiotês, (term mentioned) dissimilarity, 1101,26*; 1102,11*
anônumos, nameless, 1102,15*
anôtatô, highest, 1049,14
anous, lacking intelligence, 1077,12*
antakolouthein, follow mutually from, 1097,30
antapodidonai, give by way of contrast, 1078,3 bis
antereistikos, involving resistance, 1046,12
anthrôpeios, human, 1102,9
anthrôpinos, human, 1066,4; 1077,6; 1102,24
anthrôpos, human being, 1050,6 et passim
person, 1103,6; 1115,15
anthupopherein, (vb.) reply, 1093,17* (as emended)
antigraphon, manuscript, 1054,28; 1056,15; 1057,6; 1078,9; 1093,6
antikeimenon, opposite, 1040,21; 1045,16; 1047,14; 1077,27.28
opposed, 1067,21
inverse, 1105,23
antilambanesthai, apprehend, 1058,4
antilegein, dispute, 1082,28
antilêpsis, apprehension, 1058,1; 1059,7.8

antiperistasis, reciprocal replacement, 1050,8
antistrephein, be convertible, 1093,27*
antistrophê, conversion, 1063,6
  antistrophê sun antithesei, contraposition (lit., conversion with substitution of the contradictory), 1040,22
  antistrophôs, conversely, 1063,4
antithesis, substitution of the contradictory (see **antistrophê**), 1040,22
aoristos, indefinite, 1071,1
apagein, draw away, 1055,19*
  reduce, 1085,6; 1113,29
apagôgê, reduction, 1042,18
apaidein, lack, 1036,19
apangellein, report, 1098,13
aparithmein, enumerate, 1081,11
apatê, deception, 1096,25
apatheia, impassivity, 1069,30
apathês, unaffected, 1067,4*; 1069,29*; 1070,3.12.14; 1071,25*
apeiros, inexperienced, 1077,19
apeiros, infinite, 1042,20 et passim
  ep'apeiron, to infinity, 1042,13 et passim
  apeiroi tôi plêthei, infinitely many, 1044,25-1047,17 passim
  infinite in multitude, 1046,6
apekhein, be far removed, 1086,19*; 1096,29*; 1114,24
apeuthunein, guide aright, 1037,28
aphairein, subtract, 1113,3
aphairesis, subtraction, 1101,12; 1115,4
aphienai, dismiss, 1071,1
aphrosunê, folly, 1077,16
apithanos, unconvincing, 1055,20; 1108,5
apoballein, lose, 1066,9.25; 1073,11*
apoblêtos, to be dismissed, 1037,4; 1040,13
apobolê, loss, 1062,12*.15.22; 1068,28*-1074,1 passim
apodeiknunai, demonstrate, 1045,14; 1048,5; 1060,22; 1072,18; 1073,7
apodeixis, demonstration, 1036,6 et passim
apodidonai, propound, 1088,16
  display, 1066,22.30
  give, 1055,6; 1109,10.26
  give in answer, 1078,5.8.9.10
  make, 1055,6
apoginesthai, go out of being, 1066,15; 1071,8
apoginôskein, give up on, 1082,26
apokathistasthai, be restored, 1080,13.18
apokhôrein, withdraw, 1067,2
  diverge from, 1063,29
apokhôrizein, separate, 1060,9
apokhrasthai, making use of, 1087,15
apoklinein, incline, 1063,28
apokrisis, excretion, 1076,25
apolambanein, gain, 1062,24; 1065,11.12.16
apoleipein, depart from, 1041,1*
  vanish, 1110,26
apolêpsis, gaining, 1077,6
apollusthai, perish, 1066,9
apologeisthai, defend oneself, 1101,24
apophaskein, deny, 1074,23
apopheugein, escape, 1077,21*
aporein, pose a (the) puzzle, pose as a puzzle, 1063,30 et passim
aporia, puzzle, 1041,28; 1093,17*
aporos, puzzling, 1039,27; 1088,15; 1109,31
aporrhein, flow away, 1061,11.12 bis.13
aporrhoia, effluence, 1056,1*
aporrhutos, subject to outflow, 1077,9*
apôsis, pushing away, 1049,18.21.23; 1050,10 bis; 1056,9
apospan, detach, 1054,13.19.23
apotelein, produce, 1097,1; 1110,26
  perform, 1066,20.26
  render, 1046,3
apôthein, push away, 1053,25
apsukhos, inanimate, 1053,18; 1058,24-1061,4 passim; 1111,31
aptaistos, flawless, 1040,15
ara, therefore, 1039,9 et passim
araros (pf. part. of **arariskein**), (with **ekhein**) be fixed, 1105,28
areskein, be (someone's) view, 1050,24
aretê, virtue, 1062,9; 1065,1-1074,17 passim; 1081,26; 1113,9.13.15
aretousthai, become virtuous, 1066,5
arithmos, number, 1088,24; 1102,3-26 passim; 1115,8
  arithmôi, in number, 1043,14 et passim
  kat'arithmon, numerically, 1043,10-1044,17 passim
arkein, suffice, 1078,22.25; 1079,2; 1090,32; 1095,29; 1114,15
  be sufficient, 1097,26
  (pass.), be content, 1102,14
arkhaios, ancient, 1082,15.16
arkhê, beginning, 1037,22; 1103,28; 1111,30
  principle, 1051,18*.20*.21*; 1102,18; 1116,9
  source, 1040,18; 1048,7; 1050,3; 1082,4
  (transliterated), 1097,2.3
  (periphrasis with **poieisthai**), begin,

1079,30
ex arkhês, original, 1045,10* et passim; originally, 1045,14; 1079,10; 1080,12.22
kat'arkhas, at the beginning, 1077,11*
tên arkhên (as adv.), to begin with, 1051,27; 1074,15; at the beginning, 1095,13
to en arkhêi aitein, beg the question, 1040,9; 1041,32
to en arkhêi lambanein, assume what [was to be proved] in the beginning (i.e. beg the question), 1042,22-3
arkhesthai, begin, 1051,11; 1063,5.7.9; 1079,26;
pothen arkhomenê, original, 1093,18*
artasthai, depend, 1037,14
arutein, draw (water), 1073,5*
asapheia, unclearness, 1107,24*
asapheian ekhein, be unclear, 1094,9
asapheian eskhe, has become unclear, 1084,30
asaphôs, in an unclear way, 1093,12
asômatos, incorporeal, 1055,28
asphalês, safe, 1060,30
asphalôs, to be on the safe side, 1090,24
asumblêtos, noncomparable, 1083,4-1101,8 passim; 1113,33.34; 1114,1
to asumblêton, noncomparability, 1099,25
asummetria, disproportion, 1067,21; 1068,6; 1071,25.28; 1072,11; 1110,19; 1113,10
asummetros, disproportionate, 1073,22; 1080,17
incommensurable, 1083,2
asummetrôs, (with ***ekhein***) disproportioned, 1067,25
atelês, imperfect [or incomplete], 1065,24; 1077,23*
athanatos, immortal, 1077,8*
athroizesthai, gather, 1075,7.8
atomos, atom, 1046,19; 1069,24
individual, 1089,18
indivisible, 1091,7; 1093,9*.15*; 1097,10; 1099,27; 1101,17; 1113,22
atonos, weak, 1076,26; 1077,19
atopos, absurd, 1064,8*.14*.19*; 1065,27; 1083,30*; 1084,3
autos, same, 1036,6 et passim
(term mentioned) 'same', 1114,29
autothen, immediately, 1057,4
auxê, growth, 1076,26; 1077,13*; 1078,29
auxein, cause increase, 1048,23*; 1061,5*-15 passim; 1111,6-24 passim; 1113,1.2
increase (trans.), 1111,8
auxêsis, increase (n.), 1043,32 et passim
increasing, 1100,3
auxesthai, increase (intrans.), 1059,28; 1100,2; 1110,17; 1111,16
be increased, 1061,5*-15 passim; 1110,11; 1111,6.8*.21; 1113,1.3
(with cognate accusative) undergo an increase, 1111,19
axiologos, worthwhile, 1110,29
axiôma, axiom, 1086,9; 1103,33; 1111,11.12; 1115,18
axios, worth, 1109,24
axioun, deem it worthwhile, 1037,18
deem it proper, 1070,13

badisis, walking, 1094,21*-1095,16 passim
badizein, walk, 1095,6
bainein, stand, 1110,5*
baros, weight, 1104,2 et passim
barus, heavy, 1050,5; 1056,17*.26*.28; 1057,2
[***bathos***] ***dia bathous***, deeply, 1064,1.2
bia, force, 1037,15; 1049,17; 1054,14.19.22; 1055,12; 1077,10*; 1112,25
biaios, forced, 1049,13.16
biazesthai, be forced, 1063,7
forcibly overpower, 1056,6*
biblion, book, 1036,3 et passim
bios, way to live, 1077,22*
blepein, look, 1095,32; 1100,34
boêtheia, (n.) help, 1079,20
boêthein, (vb.) help, 1080,7
boulesthai, wish, 1074,11; 1075,5; 1085,18; 1088,3; 1100,25; 1102,2; 1107,4
want, 1095,1
would have it, 1079,12; 1108,3
bradus, slow, 1082,9-1084,27 passim; 1095,14; 1099,13; 1102,4; 1113,21; (term mentioned) 'slow', 1085,22
to bradu, slowness, 1086,1.10; 1098,16; 1113,34
bradeôs, slowly, 1098,23*; 1099,13; 1102,9; ***(to) bradeôs***, (term mentioned) 'slowly', 1085,21
brakhus, small, 1046,20

dasunein, aspirate, 1099,21
deiknunai, prove, 1037,18 et passim
point out, 1043,18
show, 1038,3; 1042,29; 1046,2; 1063,18
deiktikos, in proof of, 1090,14
deiktikôs, directly, 1047,16
deilia, cowardice, 1044,10

dein, ought, 1087,13; 1112,14
must, 1100,13; 1114,7
dei, one ought, 1073,2 bis.3.5*; 1099,3.25; 1100,27.32*; one must, 1074,22; must, 1091,5; 1092,12; 1100,13; 1114,7; have to, 1092,26; ***ouden edei***, there was no need, 1093,11
deisthai, be in need of, 1079,16.20; need, 1095,32; there is need, 1060,23
deon, needful, 1096,4
deixis, proof, 1042,18.25
dekatalantos, ten-talent, 1104,9.14
dekhesthai, receive, 1062,18; 1089,12.20; 1090,8.13.27 bis.30; 1101,4; 1114,7.9
to dekhomenon, recipient, 1114,12
to dedegmenon, recipient, 1089,15
dektikos, capable of receiving, 1089,23-1091,8 passim; 1096,2*; 1097,30; 1098,3
to dektikon, recipient, 1098,1
dêlos, obvious, 1038,15 et passim
dêloun, indicate, 1050,12 et passim
dêmiourgia, creation, 1077,7; 1102,25
deuterôs, secondarily, 1090,25.27
dexios, on the right, 1068,12
diadekhesthai, receive in succession, 1050,7
diadokhê, succession, 1046,16
diairein, divide, 1039,4 et passim
distinguish, 1037,23; 1096,5; 1099,25
detach, 1109,8.29.32; 1110,9
diairesis, division, 1041,27; 1099,27
ek diaireseôs, by division, 1072,21
diairetos, divisible, 1039,1.2; 1041,1-1042,2 passim
diakosmein, set in order, 1077,7
diakoptein, interrupt, 1043,26
diakrinein, segregate, 1052,20
dilate, 1059,4 bis
distinguish, 1085,20; 1113,33
diakrisis, segregating, 1050,19-1053,2 passim
dilating, 1073,18
distinction, 1098,11
diakritikos, segregative, 1050,15; 1051,26*
distinguishing, 1085,5
dialambanein, divide, 1049,11
dialeipein, leave a gap, 1043,22
dialuein, resolve, 1047,11
diametros, diagonal, 1083,2
dianoêtikos, mental, 1074,4
dianoia, mind, 1076,24; 1077,3*; 1078,7*
diaphanês, transparent, 1060,16*
diapherein, differ, 1044,8 et passim
make a difference, 1046,18; 1055,14.21
diaphônein, be at odds, 1095,34
diaphora, difference, 1036,5 et passim
differentia, 1094,22*; 1096,3*.5.6
variety, 1049,14.18; 1079,14
see under ***ekhein***
diaphoros, different, 1090,5-1099,29 passim; 1114,12.14
in different forms, 1086,21
diaporein, be thoroughly puzzled, 1101,16
diaporeuesthai, pass through, 1077,22*
diapseudein, deceive, 1098,15
diarthrôsis, detailed articulation, 1115,1
diarthroun, articulate, 1037,2; 1040,14; 1091,18; 1092,2; 1100,10
emended to ***diorthoun***, 1038,23
diasôizein, preserve, 1103,17
diastasis, interval, 1094,8*
diastêma, distance, 1085,13; 1094,17; 1095,19; 1098,27; 1103,8-1110,22 passim; 1115,11-1116,10 passim
diatattesthai, put (something a certain way), 1052,28
diathesis, condition, 1058,3.7; 1081,11
diatithenai, dispose, 1064,2; 1067,22*; 1071,22*.24.26
didaskalia, teaching, 1080,6.12
didaskein, teach, 1037,28; 1047,24; 1068,29; 1073,4; 1099,29; 1114,15
dielenkhein, refute, 1105,27
dierkhesthai, traverse, 1084,21*
diienai, go through, 1084,29*
dikaiôs, justly, 1038,5
dikaiosunê, justice, 1074,18
dikhôs, in two versions, 1036,4
dinein, turn (something), 1037,19 et passim
dinêsis, turning, 1052,16-1053,27 passim; 1112,27
diorismos, distinction, 1088,16; 1095,9; 1114,6
diorizein, distinguish, 1037,20; 1098,21.32
make a distinction, 1064,21
define, 1091,27
specify, 1070,26
diôrismenos, distinct, 1043,14
diorthoun, (vb.) correct, 1038,23
diôsis, pushing apart, 1050,9-1052,26 passim
diôthein, push apart, 1050,28 bis
diplasios, double, 1068,12 et passim
(term mentioned) 'double', 1087,20; 1089,13
diplasiôn (different form of adj.), double, 1111,16
to diplasion, doubleness, 1087,19-1089,30 passim; 1090,21;

1091,10; 1097,5; 1113,35; 1114,4.23
dittos, of two kinds, 1048,7; 1066,10
dokein, seem, 1036,11 et passim
dran, do, 1058,17; 1112,32
drattesthai, lay hold of, 1063,31
dromikos, capable of running, 1062,20
dunamis, power, 1050,7 et passim
force, 1103,4-1116,7 passim
potentiality, 1076,20
capacity, 1057,28; 1058,31; 1059,1; 1062,20; 1079,24
dunamei, potentially, 1075,16; 1109,10.26.27
to dunamei, the potential (as n. and adj.), 1074,21; 1075,3.23
to kata dunamin, the potential (as adj.) 1074,27*
dunasthai, be possible, 1037,29; 1097,6
can, 1039,22 et passim
be able, 1047,21 et passim
be able to do, 1087,7
dunatos, possible, 1041,4 et passim
hôs dunaton, as much as possible, 1037,9
kata to dunaton, to the maximum extent possible, 1111,28
duskrasia, bad temperament, 1080,20

edaphos, ground, 1084,7
egeirein, (pf.) be awake, 1079,22
(aor. pass.), wake up, 1080,2*
[***eidenai***] ***oida***, I know, 1052,4
isteon, one ought to know, 1036,8; 1042,7; 1051,7.9; 1054,27; 1082,13; 1086,20
eidopoiein, constitute (something's) form, 1081,30
(mid.-pass.), owe its form to, 1089,29
eidopoios, constitutive of form, 1058,3; 1081,25
eidos, form, 1056,24* et passim
species, 1043,27 et passim
eikôn, image, 1096,30
eikotôs, reasonably, 1041,5; 1045,7; 1079,4; 1083,24
einai, be, exist, be possible, 1036,2* et passim
(fut. ptc.), future, 1078,15
to einai, essence, 1069,10; being, 1071,3
to X einai, the essence of X, 1043,16; 1068,17; 1087,8.25
to ti ên einai, the essence, 1081,27
on, (subst.) being, 1100,4
ontôs, genuinely, 1037,5; really, 1079,11
tôi onti, in reality, 1094,6*
eipein, state, 1041,6 et passim
speak, 1043,28; 1056,5*
speak of, 1058,26; 1100,1; 1114,27
say, 1044,6; 1047,8; 1061,2; 1072,7; 1073,17; 1075,6; 1099,9; 1112,12
suggest, 1037,1
declare, 1089,11
eipe (imperative), tell, 1108,19*
[***eirein***], state, 1041,22 et passim
rhêtheis, stated, 1051,12 et passim; alleged, 1089,9
rhêthêsetai, will be described as, 1092,3
rhêthêsomenos, which will be discussed, 1057,6; (neut. with ***to***) what will be stated, 1086,19
eirêsthai, to have been stated, to be stated, 1041,22 et passim; to have been discussed, 1044,14.16; (of a statement) was made, 1097,7; 1106,13
to eirêmenon, statement, 1106,13
erein (fut.), will say, 1090,9; will speak, 1094,5*
rhêteon, one should state, 1054,18; one must state, 1099,4; 1102,3; 1110,6; one ought to talk about, 1114,28
eispnoê, inhaling, 1050,16; 1051,23*
ekballein, get rid of, 1114,11
ekhein, have, 1036,5 et passim
possess, 1070,6; 1071,18.24*; 1074,17; 1079,4.11; 1081,1; ***to ekhon***, the possessor, 1066,12; 1069,28*; 1071,27
bear, 1105,23
involve, 1059,9.12; 1110,29
with ***diaphora***, display, 1094,6*; 1095,33; 1096,1; 1097,27
be, 1050,4; 1066,17.19; 1079,10; 1088,16; 1091,19.28; 1095,31; 1105,28; 1110,23
ekhein pros, be related to, 1106,14
ekhei to anankaion, is necessary, 1104,23
see phrases under ***asapheia***
to ekhomenon, the next, 1043,7* bis
ekkeisthai, be set forth, 1106,23
ekkoptein, eradicate, 1070,1
ekkritikos, eliminative, 1051,25*
ekluesthai, be exhausted, 1049,26; 1050,1
ekphansis, manifestation, 1079,20
ekpnoê, exhaling, 1050,15; 1051,24*
ekstasis, departure, 1065,25
ektasis, stretching out, 1048,11
ekteinein, extend, 1098,19
ekthesis, setting out an illustration, 1084,20
ekthetikos, illustrative, 1038,2
ektithesthai, set (something) out, 1063,13; 1073,7
set out as an illustration, 1039,3

elakhistos, least, 1076,8 et passim
elattôn, less, 1076,10 et passim
fewer, 1058,14 et passim
smaller, 1061,11 et passim
minor (premise), 1041,17; 1063,14
elattousthai, be less, 1087,9(not translated).10; 1105,26; 1106,16
êlektron, amber, 1055,10
elenkhein, refute, 1085,4
reject, 1081,16
elleipein, be missing, 1093,5
elleipsis, deficiency, 1067,22
ellipês, defective, 1078,7
elpis, expectation, 1072,24
elpizein, expect, 1072,24*.27*
empeiria, experience, 1074,29
emplekesthai, be entangled, 1056,3*
empodizein, be a hindrance, 1038,15
hinder, 1078,22; 1079,4.21; 1080,3
empoiein, produce in, 1058,4.6; 1081,29
empsukhos, animate, 1053,15.16; 1058,23-1061,3 passim; 1067,2; 1111,31; 1112,23
enallax, alternately, 1104,20.24.31; 1105,4.21
enantiôs, in the contrary way, 1067,4*
enargês, evident, 1040,21; 1061,9; 1062,6.7; 1074,16; 1079,29; 1103,20.25
enargôs, evidently, 1112,23
enargesteron, more evidently, 1088,13
endein, bind in, 1077,8*.9*.12*
endekhesthai, be possible, 1045,22; 1047,3*-28 passim; 1070,27*; 1107,13; 1112,13.15
hôs endekhomenos, as a possibility, 1047,20.25.26.27; 1112,13
endidonai, supply, 1048,12
endothen, from within, 1059,8; 1080,11.14
eneinai, be present, 1076,18; 1078,24
energeia, activity, 1059,1 et passim
energeiai, actually, 1042,4 bis; 1075,16; 1109,10.12.13.26; in actuality, 1074,22.27; 1075,3
kat'energeian, in actuality, actually, actual, 1041,27; 1058,31; 1059,2*.6; 1060,16*; 1075,15; 1079,31
energein, be active, 1058,18; 1075,22*-1076,21 passim; 1079,25.28 bis; 1080,6; 1112,33
energês, (adv.) effectively, 1042,9
energêtikos, active, 1059,8
enginesthai, arise, 1058,5.7; 1069,20-1073,21 passim; 1078,23; 1090,24
enistasthai, object, 1090,1
raise an objection, 1086,29; 1113,27.35; 1114,10
enkalein, find fault with, 1040,9
ennoein, think of, 1072,29
ennoia, purport, 1037,9; 1086,25
epi tês ennoias, in mind, 1098,18
enokhlein, annoy, 1080,21
enstasis, objection, 1047,11; 1054,30; 1113,28
enteuthen, starting here, 1082,7
entreptikôs, damningly, 1090,2
entunkhanein, read, 1040,16
enuparkhein, inhere, 1114,12
eoikein, seem, 1058,26 et passim
epagein, add, 1040,26 et passim
infer, 1047,18; 1048,1; 1063,18
introduce, 1079,28; 1085,5; 1105,9; 1115,24
[**epagôgê**, induction]
apo tês epagôgês, by induction, 1048,4
dia tês epagôgês, through induction, 1057,11
ek tês epagôgês, based on induction, 1057,19.21
epakolouthein, follow after, 1049,22.24
epaktikôs, inductively, 1057,23
epanienai, return, 1059,30; 1091,22
epaphê, touching, 1047,2
epeisaktos, extrinsic, 1066,14
epeisodiôdês, adventitious, 1064,3
epekhein, hold in check, 1079,23
epeoikein, resemble, 1102,22*
ephaptesthai, touch, 1050,5
epharmottein, coincide, 1040,31; 1041,2
ephelkesthai, pull along, 1054,9
ephexês, next, 1038,18 et passim
subsequent, 1037,14
immediately after, 1056,18
ta ephexês, what comes next, 1093,2
ephistanein, remark, 1045,13; 1055,7; 1056,7; 1070,19; 1090,32; 1114,3
note, 1098,32; 1114,13
attack, 1112,5
ephodos, inspection, 1074,30
epibatês, passenger, 1053,20
epibolê, view, 1052,3
epidekhesthai, admit, 1065,23
epienai, go over, 1050,17
come on, 1077,13*
go on, 1077,14*
epiginesthai, supervene, 1069,3.5.6.11.16; 1073,14.20; 1113,12
epigraphein, label, 1036,4
epikeisthai, lie on, 1053,22
epikheirêma, argument, 1038,16; 1090,14
epikratein, prevail over, prevail, 1049,25.27.30

epinoein, think up, 1108,6
epiphaneia, surface, 1064,1; 1089,20-1094,4* passim; 1098,3.8; 1100,30; 1114,10.15.32
epipherein, draw [sc. a conclusion], 1063,16; 1082,22
add, 1089,2; 1093,1; 1094,28; 1106,5
adduce, 1097,16
epipolaios, superficial, 1058,3.7
epipolaiôs, superficially, 1064,5
epirrhutos, subject to inflow, 1077,8*
episkopein, examine, 1082,6
epistasthai, know, 1073,23*; 1075,5*; 1077,3; 1078,26; 1079,2.30
epistêmê, knowledge, 1062,9; 1074,5-1081,8 passim
epistêmôn, knowing, 1074,12; 1080,4
(as subst.) knower, 1074,12.16.27*; 1075,3.23; 1076,2; 1078,15-1079,33 passim
epistêmonikôs, scientifically, 1074,30
epistêtos, knowable, 1074,19; 1076,1; 1079,30
episumbainein, supervene, 1066,14
epitêdeios, suited, 1079,1
epitêdeiotês, suitability, 1062,20
epiteleisthai, be accomplished, 1043,4; 1054,26; 1059,6; 1076,3; 1081,3; 1102,24
epithesis, putting on, 1065,28
epithumia, appetite, 1070,2; 1071,20
epitrekhein, summarize, 1111,29
epitrepein, entrust, 1098,14
epizêtein, inquire further into, 1103,2; 1115,12
epôsis, pushing along, 1049,18.20.23
epôthein, push along, 1049,21
erastos, beloved, 1046,13
êremein, be at rest, 1038,21* et passim
êremia, rest, 1042,2; 1076,28
êremêsis, coming to rest, 1076,28
êremisis, coming to rest, 1079,3
êremizesthai, come to rest, 1076,23; 1077,25 bis-1078,17 passim; 1080,1*
eresthai, pose (sc. an argument by questioning), 1108,19
erkhesthai, come, 1055,23*; 1075,24; 1077,23*; 1103,33; 1115,5
erôtan, pose the (a) question, 1100,26; 1101,16
propound (sc. an argument by questioning), 1108,28
eskhatos, last, 1056,20*-1061,21 passim; 1066,11
ultimate, 1092,1.5.22; 1096,17; 1114,16.21
ethein, make a habit of, 1054,15
êthikos, moral, 1070,1-1074,17 passim; 1113,15
ethos, custom, 1036,4; 1044,1
habit, 1080,6.12
etumologein, derive (a word), 1077,4
euapoblêtos, easily lost, 1064,3
eudaimonein, be happy, 1073,6*
euexia, good condition, 1067,11
eulogôs, with good reason, 1089,1
eumetablêtos, easily changeable, 1058,8
eumetabolos, easily changed, 1067,26
eupathês, easily affected, 1067,15.26
euphuôs, (with ***ekhein***) be naturally well-suited, 1050,4
euthêmoneuein, be in good condition, 1067,24
euthuphoria, rectilinear locomotion, 1088,5
euthus, straight, 1082,23 et passim
(adv.) immediately, 1054,30; 1083,32*; 1095,23; 1099,19; 1115,28; right, 1111,30
eutheia, straight line, 1082,20-1098,17 passim; 1113,24-31 passim
eutreptos, easily mutable, 1067,14
exairesis, removal, 1078,22
exaptesthai, be kindled, 1056,6*
exêgeisthai, explain, 1048,17; 1051,27; 1055,16.21; 1075,4.13
exêgêsis, explanation, 1093,13; 1108,6; 1111,28
exêgêtês, commentator, 1036,7
exetasis, examination, 1040,16
exomoioun, assimilate, 1064,11
exôthen, from outside, 1037,15 tris et passim

galênê, calm, 1077,14*
calming, 1078,25
gê, earth, 1109,32; 1110,5* (Doric)
genesis, coming to be, 1043,31 et passim
genikos, generic, 1052,4; 1053,1*
gennan, generate, 1056,27
gennêtikos, generative, 1056,17*.28; 1057,2
genos, genus, 1043,30 et passim
kind, 1081,12*
geusis, taste, 1060,22; 1090,18
organ of taste, 1060,22
geustos, tastable, 1060,23
ginesthai, come to be, 1042,4 et passim
come into being, 1067,24
become, 1047,1 et passim
occur, 1042,20 et passim
(with ***en***) arrive at, 1036,16; 1077,28

be produced, 1049,13.16.17; 1072,26.27 bis
be made, 1054,16; 1114,27
ginôskein, know, 1038,17; 1075,17.19.24; 1096,4
be acquainted with, 1051,7.10; 1057,27
glukus, sweet
(term mentioned) 'sweet', 1090,9
to gluku, sweetness, 1090,10; 1098,7*.9.19
glukutês, sweetness, 1058,2
gnôrimos, known, 1057,11; 1062,22
well-known, 1101,10
gnôrizein, know, 1059,24; 1074,30; 1099,26
gnôsis, cognition, 1074,19.28.29; 1075,5.7.12.19
gnôstos, cognizable, 1074,19; 1079,25.28
grammatikos, grammarian, 1079,21
grammê, line, 1082,23-1097,10 passim; 1113,28.30
graphê, text, 1078,6; 1086,20.23; 1093,6.8.10; 1106,1.4
graphein, write, 1037,1 et passim
grapheion, pencil, 1085,26*

Haidês, Hades, 1077,23*
hairesthai, choose, 1039,16
hama, simultaneously, 1040,2 et passim
simultaneous, 1043,5 et passim
together, 1048,6 et passim
at the same time, 1051,14* et passim
hapas, absolutely all, absolutely every, 1037,11*; 1044,24*; 1051,10*.15*; 1052,10*; 1059,31*; 1065,22; 1067,3*; 1071,2*; 1073,6*; 1085,15*; 1086,21*.22*.24*; 1088,2; 1102,24
haphê, touch, 1060,3.9
touching, 1076,7.12
contact, 1046,15.17
haploïkos, simple, 1075,13
haplôs, simply, 1050,17; 1052,26; 1098,26; 1099,9
absolutely, 1050,5
haptesthai, touch, 1046,11 et passim
haptos, tangible, 1060,8
harmottein, fit, 1102,16
hêdein, please, 1072,24
(mid.-pass.) be pleased, 1072,22.28; 1073,1
hêdonê, pleasure, 1072,17*-1074,2 passim
hêdus, pleasant, 1072,23.24
hêgeisthai, suppose, 1096,25; 1097,5
lead, 1097,3
heis, one, single, 1043,10 tris et passim; (term mentioned) 'one', 1088,28.29
to hen, unity, 1043,29
aph'henos, derived from a single thing, 1096,31
pros hen, relative to a single thing, 1096,31; (in descriptions of ratios) to one (i.e. as in 'two to one'), 1086,26*; 1087,21.27; 1088,27
hekastos, each, 1043,8 et passim
to kath'hekaston, individual, 1050,17; 1074,29.30
hekatontaplasios, a hundred times, 1115,33
hekatostos, hundredth, 1103,7 et passim
hêlikia, stage of life, 1076,23
helkein, pull, 1037,19 et passim
to helkon, (the) puller, 1054,8-1055,22 passim
helxis, pulling, 1050,9-1056,12 passim; 1112,27.29
hêmiôrion, half-hour, 1104,10-1105,6 passim
hêmistadion, half a stade, 1104,15; 1105,3.4.6
[***heneka*** or ***heneken***]
to hou heneka, that for the sake of which, 1047,30*; 1082,4
to hou heneken, that for the sake of which, 1048,8
henousthai, be united, 1046,15; 1061,10
hepesthai, be consequent (upon), 1042,20 et passim
follow, 1073,13; 1083,5
follow from, 1085,3.11
Herakleios, Heraclean (see under ***lithos***), 1055,9.25
hêsukhazein, (mid.) quiet down, 1076,24
heterophuês, of different natures, 1096,26
heteros, different, 1043,21 et passim
of another kind, 1093,5*; 1097,20*.23*.24.26; 1098,1.2; 1101,2.20*.23*; 1102,17; (term mentioned) 'of another kind', 1101,24*; 1102,16.17; 1114,29
other kind of, 1062,7
heterotês, otherness, 1100,10
(term mentioned) 'otherness', 1101,26*; 1102,1*
hetoimos, prepared, 1079,23
prepared for, 1080,22
hêttôn, less, 1085,16* et passim
(neut.) to a lesser degree, 1058,14
heuriskein, find, 1051,8 et passim
discover, 1082,27-1083,2 passim
hexês, next, 1042,16
and so on, 1063,15
ta hexês, what follows, 1051,13; what

comes next, 1110,29
hexis, state, 1057,27; 1062,8-1074,20 passim; 1078,14-1081,19 passim; 1113,8.9.12.14
hikanos, adequate, 1098,12
hippos, horse, 1053,15; 1071,17 bis; 1089,16.18.21; 1090,28; 1097,25; 1098,8; 1100,18; 1114,8
hiptasthai, fly, 1095,6
histanai, (pf. act.) stand, 1084,6
histasthai, come to a halt, 1039,18.24.25; 1047,14
stênai, to have come to a halt, 1039,12.26; 1077,3*; 1078,7*.10*
hodos, way, 1077,14*
hodôi, methodically, 1092,2
holkê, drawing in, 1050,16
pulling, 1054,17 bis.26; 1055,14
holoklêros, perfect, 1077,21*
holos, whole, 1036,10 et passim
as a whole, 1037,28 et passim
universal, 1102,25
holôs, in general, 1037,21 et passim; speaking generally, 1048,9; 1066,13; 1070,27; 1082,14; at all, 1040,4 et passim
homoeidês, of the same species, 1044,10; 1082,9.19; 1091,25-1102,6* passim; 1113,22; 1115,3.8
homogenês, of the same genus, 1044,11; 1091,6; 1096,16.19.24; 1100,9; 1114,19
homoiomerês, homoeomerous, 1067,18; 1069,24
homoios, similar, 1054,20 et passim
(term mentioned) 'similar', 1114,29
to homoion, similarity, 1083,13
homoiôs, similarly, 1051,24* et passim; equally, 1047,10; 1078,27*; to a similar degree, 1099,12*
homoiôs ekhein, the situation is similar, 1091,27; be similar, 1095,31
homoiotês, similarity, 1086,20*; 1096,30*; 1097,2.18; 1098,31*; 1099,1
homologein, agree, 1038,8; 1057,25; 1062,6
ta homologoumenôs homônuma, things which are generally agreed to be homonymous, 1090,3
homônumia, homonymous meaning, 1086,19*; 1096,12*; 1097,16
homônumos, homonymous, 1085,15*-1097,18 passim; 1114,5-24 passim; homonymously, 1113,33.34
homônumôs, homonymously, 1063,10* et passim
homophuês, of like nature, 1096,23; 1097,4.11.18; 1098,17
homotakhês, of the same speed, 1082,21; 1091,16*; 1094,16
hôra, hour, 1104,9 et passim
horan, see, 1038,7; 1048,4*; 1059,2.26; 1076,9.10; 1107,20
horatos, visible, 1060,13 bis.14; 1062,3
horismos, definition, 1054,4.5 bis; 1065,24; 1085,11; 1112,29; 1114,6
horizesthai, define, 1049,19; 1084,23
be determined, 1043,15.17
hôrismenos, definite, 1102,23;
hôrismenôs, definitely, 1107,23*
hormê, impulse, 1037,17
horos, standard, 1098,6*.10
limit, 1110,9.13
hudôr, water, 1081,23; 1086,30-1087,32 passim; 1114,1.2.5
hugeia, health, 1062,9; 1065,2-1070,10 passim; 1099,7 bis.9.11; 1100,16
hugiainein, be healthy, 1069,8; 1079,22
hugiansis, getting healthy, 1099,7; 1100,17
hugiazesthai, become healthy, 1098,24*; 1099,8.16; 1100,15*
hugiês, healthy, 1077,21*
(logic) valid, 1039,24
hugrainesthai, be moistened, 1059,16; 1060,27
hugros, liquid, 1063,3.4.11
wet, 1069,21*; 1080,15
hugrotês, wetness, 1058,2; 1067,13.25; 1080,17; 1081,22.24
hulê, matter, 1059,8; 1063,11; 1064,22; 1069,18*; 1081,21
hupagein, subsume under, 1050,9; 1051,7; 1052,2.29
hupantan, meet with, 1109,31
huparkhein, pertain, 1041,31; 1063,29; 1071,23
belong, 1074,3; 1079,18; 1099,12; 1114,14
be the case, 1047,4*
be, 1062,23
exist, 1076,1; 1114,14
ta huparkhonta, properties, 1082,7
huparxis, existence, 1098,15
huperbolê, excess, 1067,22
huperekhein, exceed, 1087,28; 1102,14
huphistasthai, undergo, 1059,17
hupnos, sleep, 1080,13
hupokeisthai, (transliterated) 1058,19*.21*
be hypothesized, 1039,8.27; 1042,29; 1044,20; 1057,24*; 1104,12
be underlying, 1064,22; 1084,9; 1100,19.29; 1114,22; ***to***

hupokeimenon, the underlying thing, 1063,7-1067,1 passim; 1081,26; 1084,1.4; 1092,6-1101,3 passim; 1114,33; the thing which was underlying, 1062,26
hupolambanein, suppose, 1062,11.17*
hupoleipesthai, be remaining, 1038,4
hupomenein, endure, 1064,23; 1076,24
abide, 1066,5
hupomimnêskein, mention, 1046,26; 1074,15
remind, 1063,22; 1080,26
bring to mind, 1076,30
suggest, 1103,26
hupomnêma, commentary, 1036,2*
huponoia, conjecture, 1040,15
hupopherein, emended to **anthupopherein**, 1093,17*
hupopsia, suspicion, 1050,18
hupopteuein, suspect, 1039,16
hupoptos, under suspicion, 1087,16
hupostasis, existence, 1071,14
hupostatikos, capable of bringing [something] into existence, 1075,12
hupothesis, hypothesis, 1039,14.17.18*; 1042,29; 1047,16-1048,1 passim; 1082,28; 1112,10.12.13
hupothetikos, hypothetical, 1039,1
hupotithesthai, hypothesize, 1039,18 et passim
huptios, flat, 1084,7
husteron, later, 1037,1

iasthai, be cured, 1098,23*
idios, distinctive, 1056,9; 1074,21; 1082,8; 1101,24; 1113,20
own (adj.), 1043,13
idiai, individually, 1061,16
idiôtês, layperson, 1096,6
idiotropia, individual type, 1073,19
ienai, go, 1047,15; 1077,14*
proceed, 1071,10; 1110,29
travel, 1077,15*
iskhuein, have the strength, 1110,18; 1111,23
hold good, 1109,30 bis
iskhuros, strong, 1055,12
iskhus, strength, 1063,19; 1067,17; 1069,5; 1070,10; 1105,7*-1107,17 passim
isos, equal, 1045,6 et passim
(term mentioned) 'equal', 1090,11; 1114,29
to ison, equality, 1083,13 bis; 1088,25; 1090,13; 1098,30*; 1101,13; 1114,28
isôs, perhaps, 1040,15 et passim
isotakhês, equal in speed, 1082,13-1085,14 passim; 1092,2*-1101,17 passim; 1115,4.5.6
isotakhôs, with equal speed, 1083,31; 1114,30
isotês, equality, 1098,31*; 1099,1

kainoprepôs, in a novel way, 1108,4
kakia, vice, 1065,2-1073,25 passim; 1113,10.13
kakos, bad, 1044,7*.8.11
kakôs, badly, 1071,23*.24*.27*
kheirôn, inferior, 1075,12
worse, 1111,29
kakunesthai, become corrupted, 1066,6; 1067,26
kalein, call, 1054,15 et passim
kaloumenos, so-called, 1057,25*; 1068,26
kallos, beauty, 1067,15; 1069,5; 1070,10
[**kalos**] **kallion**, better, 1056,1; 1108,10
kalôs, rightly, 1042,22; 1045,13; 1047,19; 1064,21; 1066,31; 1098,32; 1112,13
nicely, 1077,3; 1081,14
well, 1055,25
kallion (adv.), better, 1055,20
kanôn, rule, 1085,19; 1113,32.35
kardia, heart, 1097,2
katabainein, descend, 1084,6
katakhônnunai, bury, 1079,19
katallêlôs, in corresponding respects, 1087,13
katamelês, neglectful of, 1077,22*
katanoêsis, consideration, 1037,5
katanôtizesthai, ignore, 1036,17
katantês, downhill, 1084,2
kataphatikos, affirmative, 1041,15
kataphronein, disdain, 1051,12
katapiptein, fall, 1108,20*.22*; 1116,3
kataskeuazein, establish, 1063,12; 1074,14
katastasis, settling down, 1076,23; 1077,10
katêgorein, (vb.) predicate, 1083,13.14; 1085,21-1089,28 passim; 1097,13; 1114,5.7
katêgoria, category, 1043,31*; 1044,1; 1086,17
predication, 1089,10
Katêgoriai, *Categories*, 1057,26; 1081,11; 1098,31
katekhein, hold down, 1056,5*
kateuthunein, straighten out, 1077,16*
katharôs, purely, 1059,8
katheudein, (vb.) be asleep, 1078,11*-1079,21 passim
kath'hauto, kath'heauto, in its own

right, 1037,20 et passim
kath'heauto, by itself, 1082,6; 1109,4
kath'heauton, by himself, 1109,7
kath'heautous, their respective, 1109,6
kathêgemôn, master, 1077,4
kathistasthai, settle down, 1077,14*; 1080,1*-15 passim
katho, insofar as, 1076,13
katholikos, general, 1052,27; 1113,31
katholou, universally, 1042,18; 1060,30; 1096,9
(adj.) universal, 1073,24*; 1075,6*.8.20.25; 1076,1
to katholou, the universal (n. and adj.), 1074,28-1075,20 passim
katiskhuein, overpower, 1054,9
katorthoun, do rightly, 1074,4
keisthai, be found, 1036,8 et passim
lie, 1084,7
be posited, 1039,11; 1042,13; 1045,21; 1057,25; 1104,4; 1107,4
(ptc. as subst.) posit, 1045,20
kenkhros, millet seed, 1108,20*-1109,22 passim; 1116,2.3 bis.4
kephalaion, main point, 1036,14.17; 1037,2; 1111,29
keramos, roof tile, 1065,28; 1066,7
keramoun, (vb.) roof with tiles, 1065,17
kêrinos, waxen, 1062,28
kerkisis, parting the warp, 1050,15; 1051,7.8*
kêros, wax, 1062,28 bis; 1063,8.10
khalkos, bronze, 1062,27-1063,10 passim; 1081,21
khalkous, brazen, 1062,27
kharakterizein, characterize, 1059,22; 1100,24
kharin, for the sake of, 1091,23
hou kharin, purpose, 1080,26
kharistiôn, kharistion (an instrument for weighing), 1110,4
kheir, hand, 1037,23; 1053,26
kheisthai, (perf.) have become molten, 1063,4
khiôn, snow, 1089,19; 1091,15
khôlos, lame, 1077,22*
khôrein, (with ***dia bathous***) penetrate, 1064,1.2
khôris, apart from, 1059,8; 1066,12.15.29; 1102,25; 1112,33
khôristikos, separative, 1050,12
khôristos, separable, 1058,18
khôristôs, separately, 1058,18
khôrizein, separate, 1054,8*.11.13.16.20.28*
(mid.-pass.) be separate, 1055,2*
kekhôrismenos, separated, 1049,10; 1058,17; 1082,17; 1109,14.15; 1112,32
khrê, must, one must, 1038,11 et passim; it is necessary, 1091,19
khrênai, it is necessary, 1100,34
ouk ekhrên, should not, 1047,28
khreia, usage, 1097,2
usefulness, 1037,4
khrêsis, using, 1075,21*; 1076,3.17; 1078,19; 1080,22
khrêsthai, use, 1036,13 et passim
khrôma, colour, 1058,3-1062,3 passim; 1090,18-1094,6* passim; 1098,5; 1114,14.18.19
complexion, 1067,17
khronos, time, 1042,20 et passim
khrôizein, (vb.) colour, 1091,14
khumos, flavour, 1060,23; 1085,25; 1086,1; 1087,1; 1090,10 bis.18; 1098,9
kinein, move (something), 1037,18 et passim
impart motion, 1038,4 et passim
set in motion, 1091,24
kinoun, mover, 1037,27 et passim
prôton kinoun, first mover, 1042,13 et passim
kineisthai, move (intrans.), 1037,7 et passim
be moved, 1037,8 et passim
kinêma, movement, 1070,2; 1077,18
kinêsis, motion, 1038,28 et passim
kinêtikos, capable of moving (something), 1060,16*; 1106,15.24; 1108,17.29; 1111,22; 1116,5
klinê, bed, 1062,27
kluein, hear, 1102,20*
koinopoiein, generalize, 1103,19
koinos, common, 1048,19*; 1049,23; 1053,22; 1082,19; 1096,32
in common, 1058,13; 1086,8; 1096,27
general, 1052,17; 1091,29; 1098,33; 1099,28; 1101,27; 1102,13.16
to koinon, the general [rubric], 1102,10.12.15
koinôs, generally, 1056,13; 1061,17; 1072,10; in a general sense, 1099,18; 1102,1; in common, 1074,2
koinotês, common feature, 1097,13; 1099,29
kôluein, prevent, 1038,13.18; 1040,15; 1093,13; 1112,5
kompazein, make a boast, 1110,5
kouphos, light, 1050,5; 1056,18*.26*.28; 1057,2
kratein, dominate, 1050,1; 1077,9* bis
kratunein, strengthen, 1065,3

kreittôn, superior, 1075,13
krinein, judge, 1078,27*; 1098,6*.17; 1100,11
krisis, judgment, 1098,14
kritêrion, criterion, 1098,10.11
kudimos, glorious, 1102,20*
kuklophoria, circular locomotion, 1088,4
kuklos, circle, 1063,8; 1065,20.21 bis; 1077,16*; 1082,21-1098,17 passim; 1113,24
kuknos, swan, 1091,15
kuôn, dog, 1089,17.18.22; 1098,8; 1114,8
kurios, important, 1036,9; 1066,28.32
  authoritative, 1054,4
  kuriôs, in the strict sense, 1056,22*; 1057,16; 1061,14; 1065,22; 1066,26

lambanein, take, 1037,26 et passim
  assume, 1039,28 et passim
  grasp, 1057,21
  acquire, 1058,5; 1065,14*; 1068,15; 1073,11*; 1077,13*; 1079,6-1080,12 passim
  to en arkhêi lambanein, assume what [was to be proved] in the beginning (i.e. beg the question), 1042,22-3
lanthanein, escape notice, 1096,20*
legein, say, 1037,23 et passim
  speak of, speak of as, 1044,1 et passim
  claim, 1038,5 et passim
  mean, 1043,20; 1046,12; 1048,17 bis; 1049,3; 1056,21*; 1057,2.13.17; 1058,21; 1062,2.3; 1080,11; 1091,9
  state, 1078,9
  (of statements) were made, 1080,26
  (with ***peri***) talk about, 1057,6; 1061,2
  kat'amphoterôn legesthai, be ambiguous, 1088,20
lêgein, leave off, 1041,1*
lêpsis, acquisition, 1062,12*.14.22; 1068,28*-1079,33 passim
lêptikos, assimilative, 1051,25*
leukainesthai, turn white, 1063,1; 1099,16; 1100,2.14*; 1114,32
  become paler, 1059,27
leukansis, turning white, 1100,3.6.17; 1114,31
leukos, white, 1043,23 et passim
  (term mentioned) 'white', 1090,6
  to leukon, (subst.) white, the white, 1088,8* et passim
leukotês, whiteness, 1083,15; 1100,15
lexis, wording, 1036,5; 1051,6; 1052,8.21.23
  text, 1054,27.31; 1056,8.15.16.22*; 1057,6
  passage, 1094,9; 1106,1.8
lithos, stone, 1049,27.29; 1054,24; 1109,32
  Hêrakleia lithos, Heraclean stone (i.e. a magnet), 1055,9.25
logikos, logical, 1036,12
  rational, 1066,24.27; 1074,5.30
  pertaining to rational disciplines, 1080,9
logos, argument, 1038,15 et passim
  account, 1052,25; 1055,6; 1069,4; 1079,12; 1088,9; 1114,11
  definition, 1087,1.2.21; 1088,13.19.20.23.27; 1089,1; 1097,1
  word, 1044,27
  reason (i.e. the faculty), 1077,19; 1080,9; 1098,15
  discussion, 1102,29; 1115,10
  ratio, 1087,21.26; 1107,3.10; 1108,1*.23*.25*
  role, 1094,1*
  sentence, 1078,7
  ekhein logon, make sense, 1054,29
loipos, remaining, 1052,16; 1053,6; 1064,13
  (n. subst.) the rest, 1066,29; 1092,9
  loipon, next, 1038,24 et passim; in the future, 1076,21.22; it remains, 1054,2
luein, solve, 1081,14
  refute, 1108,18.29; 1113,28; 1116,2.4
lupê, pain, 1072,17*-1074,2 passim
lupein, cause pain, 1072,25
  (mid.-pass.) be pained, 1072,22.28; 1073,2
lupêros, painful, 1072,23.25

makaros, blessed, 1102,20*
mallon, rather, 1041,3 et passim
  more, 1055,9 et passim
  to a greater degree, 1058,14
  ek tou mallon, on the basis of an *a fortiori* argument, 1074,14
malthakos, weak, 1036,12
manousthai, be rarefied, 1069,19*
manthanein, learn, 1042,17; 1078,27*; 1079,2.16
  mathêsis, learning, 1079,11.16
matên, in vain, 1095,34
medimnos, medimnus, 1108,21*-1109,2 passim; 1116,2
megas, great, 1037,4
  meizôn, greater, 1045,9* et passim; larger, 1101,7*.28; 1102,13.15; major (premise), 1063,15
  megistos, greatest, 1077,21*; 1110,11
megethos, magnitude, 1044,25 et passim
meiôsis, decrease (n.), 1043,32 et passim
meioun, cause decrease, 1061,8.13.15; 1113,2.3
meiousthai, decrease (intrans.), 1059,28; 1110,16

be decreased, 1061,9.11.12.14.15; 1113,2.4
mêkos, length, 1083,10 et passim
mêkunein, prolong, 1050,18
melainesthai, turn black, 1063,1
become darker, 1059,28
melansis, turning black, 1100,6
melas, black, 1043,23 et passim
menein, remain, 1050,28 et passim
remain stationary, 1055,14.21*.22*; 1056,5*
merikos, particular, 1076,1
to merikon, (n.) the particular, 1075,16.19.20.23.24
merizein, partition, 1041,30; 1104,33; 1105,9; 1109,13; 1115,24; 1116,6
meros, part, 1038,3 et passim
to en merei, the particular, 1073,24*; 1075,6*.7.9.15.17.20
to kata meros, the particular, 1075,2.4.10; 1098,10
mesos, intermediate, 1063,26
middle, 1066,11
meson (subst.), intermediary, 1057,16; medium, 1060,12.23
ana meson, in the middle of, 1061,6*.8*.15.16.19*
en mesôi, in between, 1110,12
metabainein, pass to, 1067,10; 1071,11
pass on to, 1101,10; 1113,13
metaballein, change (intr.), 1043,23 et passim
change (tr.), 1063,31
cause change, 1056,23*.25*; 1057,14
metaballesthai, be changed, 1056,23*.25*
metabasis, passage, 1064,6
metabolê, (n.) change, 1043,16 et passim
metakeisthai, be shifted, 1057,5
metalambanein, substitute, 1041,15
metalêpsis, substitution, 1055,20
metaskhêmatisis, reshaping, 1063,19; 1064,13
metatithenai, transfer, 1086,24; 1093,11
metaxu, between, in between, 1048,12 et passim
metekhein, partake, partake of, 1058,6.14; 1068,11.18.19; 1081,18.29; 1091,14
meterkhesthai, proceed, 1036,15
methê, drunkenness, 1080,13.19; 1081,4
metheuein, be drunk, 1078,11*-1080,16 passim
methienai, set free, 1073,5*
methodos, method, 1099,28
metron, (n.) measure, 1082,19; 1088,23; 1098,28
due measure, 1073,15
metroun, (vb.) measure, 1111,9
moderate, 1070,2.4
memetrêmenos, (adj.) moderate, 1070,12.14
memetrêmenôs, moderately, 1069,30
mimnêskein, (mid.-pass.) remember, 1072,22*.26*.28*; 1074,23
mimousthai, imitate, 1102,25
mnêmê, memory, 1072,23
mnêmoneuein, mention, 1098,19.22
mokhtheros, bad, 1069,29; 1070,11.13
monas, unit, 1097,3
monos, only, 1036,5 et passim
monôs, only, 1068,25
morion, portion, 1039,7 et passim
part, 1037,22; 1058,29; 1059,3; 1061,4; 1099,9; 1102,8; 1115,8
morphê, figure, 1062,10*; 1064,28; 1069,12*.13*.17*; 1081,13*.28
mousikê, music, 1091,3
mulê, millstone (in a handmill), 1053,26
mulos, millstone, 1053,21
muriostos, ten-thousandth, 1108,21*-1109,21 passim; 1116,4

naus, ship, 1053,19 et passim
nekros, dead, 1066,23.29
(subst.) corpse, 1066,24
nekroun, deaden, 1066,27
neôlkein, haul (a ship), 1107,18; 1115,30
neôlkos, ship-hauler, 1107,20; 1109,7.22; 1115,32
neos, (subst.) young, 1077,18
recent, 1082,16
nêphein, be sober, 1079,22; 1080,1*.17.21; 1081,5
nêtê, high note, 1085,27*.28; 1086,11*
neuron, sinew, 1058,24; 1067,18
noêma, thought, 1040,13
noêsis, thinking, 1037,27
noêtikos, intellectual, 1072,13-1075,3 passim; 1113,15
nomizein, believe, 1037,28 et passim
nosein, be sick, 1066,21; 1078,19.21; 1079,3.8
nosos, sickness, 1062,10; 1065,2.8; 1070,5; 1077,22*; 1080,13
nous, intellect, 1059,8 (as emended); 1075,12.14.16; 1078,24; 1081,2
nun, now, 1036,7 et passim
here, 1046,21 et passim
present, 1095,9
nuxis, spur, 1079,17

oiesthai, think, 1037,3 et passim
oikeios, own proper, 1049,28; 1051,1;

1065,11 bis.18.19; 1066,18 bis.19
proper, 1049,26; 1054,10.13.22; 1066,30; 1067,23* bis.27; 1071,22*.24*.26*
appropriate, 1059,6; 1100,28; 1102,3
germane, 1036,10; 1037,3; 1102,29; 1115,9
oikia, house, 1064,14.18; 1065,28; 1066,8.17.31
oikos, house, 1065,17
okhein, carry, 1037,19 et passim
to okhoun (ptc.), (the) carrier, 1053,11.13.14.18
okhêsis, carrying, 1052,16; 1053,3*.6.7.9.22.23; 1112,27
onkos, volume, 1085,25-1091,11 passim; 1114,3
onoma, term, 1048,16 et passim
onomazein, name (vb.), 1073,19; 1098,18; 1102,11.14; 1107,4
ontôs, see under ***einai***
onux, nail, 1058,25
opheilein, be obliged, 1048,3
ophthalmos, eye, 1099,9
opsis, view, 1057,22; 1059,31
sight [i.e. organ of], 1060,15*.17
optikos, visual, 1060,15
orektikos, desiderative, 1080,8
orekton, desired, 1048,10
orexis, desire, 1048,10; 1077,17; 1080,6
organon, organ, 1095,2
instrument, 1110,3
orthos, correct, 1077,21*
upright, 1084,6
orthôs ekhein, be correct, 1088,16
ouk orthôs legein, be incorrect, 1090,15
osphrêsis, organ of smell, 1060,17
smell, 1060,21
ôsis, pushing, 1049,18-1056,12 passim; 1112,27.29
ostoun, bone, 1058,25; 1067,18
ôthein, push, 1037,18 et passim
oudamou, nowhere, 1041,3
ouriodromein, run with a fair wind, 1095,6
ousia, substance, 1043,31* et passim
essence, 1079,15; 1081,17; 1096,27; cf. 1064,15
ousiôdês, essential, 1081,25
substantial, 1058,8.31; 1059,10
ousioun, invest (something) with essence, 1081,29
oxos, vinegar, 1085,26
oxus, sharp, 1085,26*.28
(term mentioned) 'sharp', 1085,29; 1086,12*
to oxu, sharpness, 1085,24 bis.25.27; 1086,30
paideusis, education process, 1077,21*
paidion, little child, 1078,26*.29; 1079,1.6.14
pais, child, 1079,18
pantakhou, everywhere, 1041,2
always, 1074,23
pantelôs, totally, 1077,21*
pantôs, absolutely, 1095,21
invariably, 1095,31; 1106,22.23; 1111,22
assuredly, 1070,23
paraballein, compare, 1066,32; 1089,19; 1090,4
set alongside, 1082,14.15; 1087,14
correlate, 1103,1.7; 1115,10
parabolê, correlation, 1103,1
comparison, 1114,27
paradeigma, example, 1038,2 et passim
paradeiknunai, point out, 1074,16
show, 1057,23
paradidonai, offer, 1037,6.9; 1099,28; 1101,9; 1113,32; 1115,1
give, 1059,13; 1082,7
paraginesthai, come to be present, 1065,7
paragraphein, append, 1093,13
parakhrôsis, tincture, 1064,3
parakolouthein, follow closely, 1036,14; 1037,9
paralambanein, employ, 1040,9; 1103,25; 1107,23*; 1108,4
paraleipein, leave out, 1050,18
parallagê, variation, 1087,15
parallaxis, variation, 1088,14
paralogismos, fallacy, 1085,4
paranêtê, next-to-the-highest note, 1085,28; 1086,11*
paraphrazein, paraphrase, 1036,16; 1051,12
paraskeuazein, prepare, 1079,24
paraskeuê, preparation, 1079,17
parateinesthai, be stretched out alongside, 1092,8
parathesis, citing instances, 1060,1
paratithenai, cite, 1113,35
(mid.) cite, 1057,20; 1065,20; 1066,16; 1086,29
paratrekhein, pass over, 1099,28; 1115,3.5
pareinai, be present, 1041,2.3; 1048,10; 1079,25.27
parekhesthai, furnish, 1050,18; 1103,17
parelkein, be redundant, 1036,11
paremballein, interject, 1093,14*
parempiptein, be inserted, 1049,11
parerkhesthai, pass by, 1036,14
parienai, omit, 1070,11; 1085,4; 1101,9.11.14; 1104,27

paristanai, defend, 1085,2
parônumiazein, refer to paronymously, 1063,29
parônumôs, paronymously, 1062,27
paskhein, be affected, 1057,26 et passim
suffer, 1059,5
undergo, 1072,27*.28
undergo an affection, 1063,2
patêr, father, 1102,20* bis
pathêtikos, affective, 1057,25*-1065,4 passim; 1068,26; 1070,7.8; 1080,27; 1081,14; 1112,31; 1113,6.17
subject to being affected, 1059,9.12; 1069,30*; 1071,25*; 1072,7
subject to, 1070,13 bis
pathos, affection, 1058,1 et passim
pauesthai, cease, 1038,25* et passim
pêgê, spring, 1073,4*; 1097,2
peira, trial, 1075,1.2
peirasthai, try, 1084,7
pelazein, be adjacent, 1058,16; 1060,12.13.18
pentetalantos, five-talent, 1104,10
peperasmenos, finite, 1042,20 et passim
peras, end, 1041,18
[***peri***] ***hoi peri***, the circle of, 1050,22
peribolê, laying round, 1065,27
peribolos, circuit wall, 1066,7
periekhein, encompass, 1075,20
to periekhon, surroundings, 1067,14.15.27
perilambanein, comprehend, 1075,9
periodos, revolution, 1077,8*.13*
periphereia, curve, 1082,26; 1084,22.25.28; 1085,3; 1091,26; 1092,19; 1096,8; 1113,28.29.31
peripherês, curved, curved line, 1082,23-1096,23 passim; 1113,25
periphora, rotation, 1077,16*
periplanasthai, go astray, 1098,10
perittos, otiose, 1036,15; 1057,4
petra, rock, 1039,19
phainesthai, appear, 1064,20; 1098,12; 1103,18
be apparent, 1037,29; 1038,9.11.13.14.17.22
(with ptc.) to be plainly, 1036,17; 1040,14; 1078,3
(with ptc.) to be shown to be, 1088,4
to phainesthai, the appearance, 1038,23
phaios, grey, 1091,13
phanai, say, 1036,12 et passim
assert, 1046,8 et passim
ask, 1101,16
phaneros, evident, 1038,1 et passim
(adv.) evidently, 1049,10
phantasia, appearance, 1096,32; 1098,14; 1103,17
phantastikos, imaginative, 1075,1.12
phaulos, wicked, 1066,6
pherein, cause locomotion, 1048,23*
carry along, 1077,10*
pheresthai, move locally, 1049,1* et passim
be borne, 1049,27
be carried along, 1077,10*; 1092,9
be transmitted, 1036,4.6; 1054,29; 1086,21; 1093,8; 1106,1
philologos, scholarly, 1039,13
philosophos, philosopher, 1036,1*
philosophôs, in philosophical fashion, 1085,19
phônê, voice, 1089,24*-1090,10 passim; 1098,9 bis
phora, locomotion, 1048,24 et passim
phortion, load, 1104,17
phôs, light, 1060,13*.14.15*
phôtizein, illumine, 1060,14
phronein, understand, 1078,23.25; 1079,2; 1081,3
phronêsis, understanding, 1079,6.15
phronimos, intelligent, 1080,4
phtheirein, destroy, 1067,1
(middle) perish, 1050,25; 1051,1; 1067,25-1073,10 passim
phthengesthai, (vb.) voice, 1077,4
phthinein, cause diminution, 1048,23*
diminish (intrans.), 1061,11
phthongos, sound, 1085,25.27 bis; 1087,2
phthora, perishing, 1043,32 et passim
phuein (perf.), be naturally constituted, 1069,28*; 1070,6; 1110,7.8.18.21.22
be naturally, 1072,7
phulattein, keep, 1054,22; 1063,9.23; 1104,33
maintain, 1051,13
phusikos, having to do with nature, 1037,14
natural, 1055,12; 1078,25*; 1080,7
Phusikê akroasis, *Physics*, 1036,2*.3 (see note on 1036,3)
(subst.) physicist, 1051,17*; 1082,16
phusikôs, naturally, 1055,25
phusiologia, doctrine of nature, 1037,5
phusis, nature, 1037,16 et passim; (not translated in 1090,2, where '***phusis*** of X' is used periphrastically for 'X')
kata phusin, in accordance with nature, natural, naturally, 1037,16 et passim
to kata phusin, [being in] accordance

with nature, 1066,21.23; 1071,10; natural condition, 1065,9.10; natural, 1080,18
phuton, plant, 1059,20; 1061,4
pikrotês, bitterness, 1058,2
piptein (with ***eis***), fall under, 1051,15*; 1052,15
pistousthai, confirm, 1048,4; 1069,26
plagios, sideways, 1049,27
epi to plagion, to the side, 1050,4
planê, error, 1097,17
plein, sail, 1039,19*; 1053,20
pleonazesthai, be greater, 1106,16
plêthos, multitude, 1046,6
see ***apeiroi tôi plêthei***
plêttein, deliver a blow, 1110,26
pleura, side, 1083,3
pneuma, pneuma, 1080,18
poiein (act. and mid.-pass.), make, 1041,16 et passim
produce, 1042,25 et passim
do, 1046,2
(middle) perform, 1040,22
(middle used periphrastically with noun for corresponding verb), 1079,30
poiêtikos, productive, 1056,16*; 1057,2
(to) poiêtikon, (the) efficient [cause], 048,8.9.14
poiêtikôs, in the manner of an efficient cause, 1048,21
poion, (subst.) quality, 1048,22 et passim
qualified thing, 1056,17*.27.28
[the category of] quality, 1068,7
[category of] quality, 1098,31*
poiotês, quality, 1043,32* et passim
polis, city, 1073,6*; 1097,3
politikos, civic, 1070,1
pollakhêi, in many ways, 1040,14
pollakhou, in many places, 1039,16; 1054,14
in many cases, 1098,15
pollaplasios, many times [sc. as great as], 1087,6; 1103,3.4; 1107,14; 1110,1; 1115,12.13
pollostêmorion, fraction, 1103,5 bis; 1110,2; 1115,13.14
polupragmonein, fuss with, 1093,11
polus, much, many, 1036,17 et passim
(compar.) several, 1044,5 et passim
great, 1076,24; 1077,9*; 1078,12*.28*; (compar.) greater, 1086,16; 1102,5*.7 bis; 1110,21 bis; 1115,7; (superl.) greatest, 1106,26; 1109,23; 1110,8
(term mentioned) 'much', 1087,3.12.17.18; 1088,21; 1089,13
to polu, muchness, 1086,29-1089,29
passim; 1090,21; 1091,9; 1097,5; 1113,35; 1114,4.23
poma, drink, 1050,16
porisma, corollary, 1096,15
poson, (subst.) quantity, 1048,22 et passim
[category of] quantity, 1098,31*
(pl.) how many, 1099,25*.26*
poson ti, of some quantity, 1103,29*.30*
posotês, quantity, 1043,32; 1058,12; 1097,15; 1101,27
potamos, river, 1077,9*
pou, where (as category label), 1044,1; 1086,7.17
pous, foot, 1094,21*.26.27
pragma, thing, 1098,16
pragmateia, treatise, 1036,9 et passim
prattein, do, 1072,22*.25*
proagein, (vb.) present, 1093,13
proballein, put forward, 1037,7
problêma, problem, 1036,5.9
prodeiknunai, prove before, 1076,30
prodêlos, obvious in advance, 1096,6
quite obvious, 1069,26
prodêlôs, obviously, 1038,1.12
proerein, speak before, 1044,12
proeirêmenos, just stated, 1045,12; just described, 1091,23; aforementioned, 1052,2; stated previously, 1101,10; (subst.) preceding statement, 1080,26
proerkhesthai, proceed, 1047,16; 1065,12; 1111,27
proelthôn, further on, 1055,21; 1094,5*
progumnazein, train in advance, 1037,4
proïenai, proceed, 1042,19; 1092,2; 1105,27; 1111,1.4.26
prokeisthai, be proposed, 1046,20; 1069,22; 1070,11
(mid.) propose, 1080,27
before us, 1091,23
present, 1036,9; 1052,17
to prokeimenon, the point proposed, 1037,13; 1038,18.24; 1055,28; 1057,21; 1059,30; 1063,17; 1065,5
prokheirizesthai, take up, 1036,8
make ready, 1079,21.22.26
prokheiros, ready to hand, 1075,15; 1079,13 tris.22.27
readily accessible, 1095,30
prokhôrein, proceed, 1109,25; 1110,5.25; 1116,13.14
prolambanein, assume in advance, 1038,25; 1040,19; 1082,21; 1103,20.33; 1115,17
precede, 1039,2; 1047,10
prolêpsis, preconception, 1038,23
proödos, procession, 1047,14; 1078,1

prophanês, manifest, 1037,18
prosagein, draw toward, 1053,25; 1055,19*
prosagoreuein, designate, 1049,4; 1063,11
prosballein, strike, 1075,16
prosdiorismos, additional specification, 1045,14
proseinai, be added, 1099,11
prosêkein, be appropriate, 1045,14
befit, 1066,25.26.28
should, 1046,5*
prosekhês, proximate (i.e. immediately next), 1057,15; 1060,21; 1061,20.21
prosekhôs, proximately, 1046,10 et passim
to prosekhôs kinoun, the proximate mover, 1048,2.6.15
proseoikein, be like, 1066,24
prosginesthai, accrue, 1078,16
prosgraphein, append [sc. in writing], 1052,3
prosiesthai, accept, 1050,9
proskeisthai, be added, 1083,21
proskrisis, accretion, 1076,25
proslambanein, assume in addition, 1039,1
proslêpsis, acquisition, 1066,7
proslogizesthai, reckon in addition, 1066,29
prosthêkê, addition, 1054,30; 1101,12; 1115,4
pros ti, relative, 1067,6-1068,22 passim; 1070,18-1075,25 passim; 1081,8 bis; 1113,10.11.14
to pros ti, the [category of] relative, 1068,1.5.10; 1071,11; 1072,11.14; 1074,7(pl.).11.14.18.20; relativity, 1071,21; 1074,17
prostithenai, add, 1040,32 et passim
protasis, premise, 1041,18; 1063,13; 1073,7
proteron, earlier, 1037,1 et passim
first, 1072,10
protithenai, put forward, 1099,6.26; 1105,26
put before, 1040,16
propose (usually mid.-pass.), 1040,17; 1042,7; 1045,15; 1048,5; 1057,10; 1091,24; 1095,13; 1102,14
prôtos, first, 1037,7 et passim
primary, 1047,30*; 1048,24; 1067,12; 1069,22*.23; 1089,12.14.23; 1090,17*.24*.25*.29.31*; 1096,3*; 1098,3; 1114,7
prôton, (adv.) first, 1037,27 et passim
to prôton, at first, 1077,12*
prôtôs, primarily, 1037,21 et passim; first, 1046,12 et passim; for the first time, 1105,9
prôtistos, very first, 1048,13
prôtotupon, original, 1096,30
psimuthion, white lead, 1089,19
psophein, make a (the) sound, 1108,22*-1110,24 passim; 1116,3 bis.6.8
psophos, sound, 1108,20*-1110,28 passim
psukhê, soul, 1038,8 et passim
psukhesthai, be cooled, 1059,16; 1060,27; 1063,1; 1068,12; 1069,2
psukhikos, psychic, 1037,17; 1065,1; 1072,1.5.10.12
psukhros, cold, 1062,29; 1069,8.21*; 1070,3; 1071,21; 1072,5; 1080,15
psukhrotês, coldness, 1058,2; 1060,11; 1067,12.24; 1081,22.24
psuxis, cooling, 1073,18
pterux, wing, 1094,26.27
ptêsis, flying, 1094,28; 1095,4.7.16
ptusis, spitting, 1050,15; 1051,24*
puknousthai, be condensed, 1069,19*
puktikos, capable of boxing, 1062,20
pur, fire, 1054,25 et passim
puraktoun, heat red-hot, 1064,5
puramis, candle, 1062,28
purios, fiery, 1064,7

rhabdos, rod, 1048,13
rhein, flow, 1073,5*
rhêsis, passage, 1051,10; 1056,7; 1105,27
rhêton, statement, 1086,20; 1094,9
rheuma, stream, 1077,13*
rhipsis, throwing, 1049,24-1050,8 passim
rhiptein, throw, 1049,25.26.29.30; 1050,3
to rhipsan, (the) thrower, 1049,30; 1050,2
rhopê, inclination, 1049,28

saphênizein, clarify, 1036,7
saphês, clear, 1100,10; 1101,14
sarx, flesh, 1058,24
sêmainein, mean, 1055,16* et passim
signify, 1091,8; 1101,25.26.27.28
sêmeion, sign, 1075,6
point, 1097,3
sidêros, iron, 1055,10; 1064,5
skepteon, one must consider, 1094,22*; 1096,3*; 1101,6*
skhêma, figure (of inference), 1041,8.11.16; 1062,24
shape, 1057,28 et passim
skhêmatisis, shaping, 1063,17
skhêmatizesthai, be shaped, 1062,16-1064,12 passim; 1081,21
skhesis, relationship, 1068,15
skhoinos, rope, 1048,13

sklêros, hard, 1063,3
skopein, consider, 1046,21; 1095,29.31
skopos, aim, 1041,22
sôizein, preserve, 1097,2; 1104,8.16; 1109,13.14.25; 1111,24.25; 1116,7.8.12.13
sôma, body, 1037,16 et passim
sômatikos, corporeal, 1055,29; 1056,2*.26*; 1112,8.16
bodily, 1046,10*.11*; 1065,2-1077,18 passim; 1113,13
sômatikôs, in a bodily way, 1046,13 bis; 1049,12; in a corporeal way, 1110,12
sophistês, Sophist, 1108,19
sôphrosunê, temperance, 1044,8.9; 1074,17
spathêsis, tamping the woof, 1050,16; 1051,6.8*
sphodros, vehement, 1077,18
sphodrotês, vehemence, 1049,25
spoudaios, excellent, 1066,6
stadiaios, of a stade, 1104,9.14.22.28.29; 1105,3
stadion, stade, 1104,10-1105,30 passim
stasis, halt, 1078,24
stathmistikos, weighing, 1110,3
sterein, deprive, 1066,24
stoikheion, element, 1056,27*; 1102,23
strephein, (mid.-pass.) twist and turn, 1040,14
strombos, top, 1053,21
sullogismos, syllogism, 1073,7
sullogizesthai, argue syllogistically, 1062,24
syllogistically deduce, 1066,21
sumbainein, result, 1044,30 et passim
happen, 1056,20*; 1057,3*
sumbebêkos, incidental, 1064,16; 1081,28; (n.) incidental feature, 1064,3; 1066,32; 1067,2
kata sumbebêkos, incidentally, 1037,21 et passim; incidental, 1070,29*; 1071,1
sumballein, compare, 1082,10.12.14; 1087,32; 1100,33*
sumblêtos, comparable, 1082,2*-1101,8 passim; 1113,20-1115,1 passim
summetria, due proportion, 1067,11-1074,18 passim; 1110,19; 1111,4; 1113,10
summetros, duly proportioned, 1068,13; 1073,21
to summetron, due proportionality, 1067,13
summetrôs (with **ekhein**), (be) duly proportioned, 1067,23
sumparalambanein, take along with, 1058,26
sumparateinesthai, be stretched out alongside, 1094,14
sumperainesthai, conclude, 1056,13; 1061,17; 1082,3; 1083,24; 1095,18; 1113,16
(with **logon**), draw the conclusion of the argument, 1047,11
sumperasma, conclusion, 1041,16; 1063,16; 1065,13.14
sumphônein, be in harmony, 1052,23
sumphônos, in harmony, 1052,24; 1077,4
sumpilêtikos, which tends to compress, 1050,13
sumplêroun, (vb.) complete, 1040,29; 1066,11.18; 1102,8
sunagein, infer, 1038,14 et passim
make the inference, 1041,11
bring together, 1074,29; 1075,2.4.11; 1082,17
formulate (an argument), 1052,17; 1077,24
sunagôgê, inference, 1041,7
conclusion, 1047,7
sunagôgos, which brings together, 1050,12
sunaidein, be consonant with, 1106,5
sunairein, put succinctly, 1116,6
sunêirêmenôs, succinctly, 1063,13; 1073,7
sunamphoteron (subst.), sum, 1109,20
sunaptein, join, 1054,14; 1093,15*
sundiairein, divide as well, 1104,12; 1115,23
sunêgoria, defence, 1085,4
suneinai, be associated with, 1049,12
sunekheia, continuity, 1051,13
sunekhês, continuous, 1043,11 et passim
sunekhizein, make continuous, 1054,12.22
sunektikos, which hold together, 1037,5
sunephelkein, pull along after, 1053,4*; 1054,17*
sunepilambanein, take part in, 1077,20*
sunethismos, habituation, 1073,1; 1074,4
sunethizein, accustom, 1037,4
sunexetazein, examine together, 1082,15
sungeneia, affinity, 1036,18
sunistanai, establish, 1044,17
(1 aor.) construct, 1045,18; 1110,4
(pf.) be constructed, 1044,28
(mid.-pass.) come to exist, 1074,5.27
(2 aor.) exist, 1102,26
sunkeisthai, be composed, 1043,12; 1046,23; 1053,8.24; 1089,1; 1102,4.10; 1114,6
sunkhôrein, admit, 1047,20.22; 1074,25; 1106,13

sunkhrasthai, make use of, 1061,26
sunkineisthai, move along with, 1038,9
sunkrinein, aggregate, 1052,20
compress, 1059,3.4
judge together, 1082,14.16
make a comparative judgment, 1085,26
compare, 1087,13; 1090,10.15.29.31; 1091,9.14.21; 1099,17; 1114,18
sunkrisis, aggregating, 1050,19-1053,1 passim
compressing, 1073,18
comparison, 1091,10.20; 1102,11.29; 1115,10
(term mentioned), 1082,16
sunkritikos, aggregative, 1050,17; 1051,26*
sunkrouein, bring into collision, 1088,9
sunônumia, synonymy, 1087,15
sunônumos, synonymous, 1086,22*; 1088,10.17; 1089,9; 1091,12
sunônumon, synonymously, 1113,34
sunônymôs, synonymously, 1085,23-1089,22 passim; 1114,1.4.7.18
sunôsis, pushing together, 1050,9-1052,26 passim
sunôthein, push together, 1050,27.29
suntattein, include, 1037,3
sunteleioun, contribute, 1066,2; 1081,17.27
sunthesis, combining, 1044,25
suntithenai, combine, 1044,24.26; 1045,28; 1109,14.16*.17; 1110,14; 1116,8
suntomos, concise, 1095,27
suntomôs, concisely, 1073,1; 1110,6
sunuparkhein, coexist, 1039,22
sustasis, constituting, 1102,24
sustoikhos, on the same level, 1075,19

takhus, fast, 1054,7*-1055,10 passim; 1082,9-1086,4 passim; 1095,5.14; 1099,13; 1101,19*-1102,8; 1113,21; (term mentioned) 'fast', 1085,22
to takhu, fastness, 1086,1-15 passim; 1088,4; 1098,16; 1113,33; ***takheôs***, quickly, 1098,23*; (compar.) faster, 1099,13; 1102,8; ***to takheôs***, (term mentioned) 'quickly', 1085,21
talanton, talent, 1104,25
tarakhê, disturbance, 1076,24; 1078,12*-1080,3 passim
tarattein, disturb, 1079,2
tattein, assign, 1040,1.8
tautotês, sameness, 1100,11.23
taxis, order, 1036,6
tekhnê, craft, 1102,25
tekhnikos, pertaining to crafts, 1080,9
tekhnologia, systematic discussion, 1060,23
tekmairesthai, judge, 1041,17
teleios, perfect, 1065,16*; 1066,22; 1071,16
(n.) adult, 1077,16; 1078,27
to teleion, perfection, 1065,21
teleiôs, completely, 1063,25; 1069,29; (compar.) in a more complete form, 1085,4
teleiôsis, perfecting, 1065,25; 1066,2 bis; 1072,11*
teleiotês, perfection, 1065,11-1071,14 passim; 1081,6.7.17.19
teleioun, (vb.) perfect, 1062,25; 1081,26
teleôs (variant form of ***teleiôs***), completely, 1063,26.30; 1069,29; 1070,1; 1079,25
teleutaios, final, 1036,10.15; 1037,2.6
[***telikos***] ***to telikon***, the final [cause], 1048,8
telos, end, 1039,2; 1040,26; 1048,9; 1110,31*; 1111,27
temnein, cut, 1103,28
tetragônismos, squaring (sc. of the circle), 1082,27.29
thaumastôs, admirably, 1065,20
thattôn, see ***takhus***
theôrein, observe, 1069,4; 1098,8
consider, 1069,31
contemplate, 1081,2
theôrêma, theorem, 1037,5.14
subject for speculation, 1115,9
theôria, theory, 1095,28
speculation, 1110,29
thermainesthai, be heated, 1059,16; 1060,26; 1062,30; 1064,5; 1068,12.16.18; 1069,1
thermansis, heating, 1068,11.18; 1073,18
thermos, hot, 1062,29; 1063,5*.12; 1069,8.20*; 1070,3; 1071,20; 1072,4; 1080,15
thermotês, hotness, 1058,1.6; 1060,11; 1067,12.24; 1080,17; 1081,22.24; 1083,15
thnêtos, mortal, 1077,7.12*
thrinkos, coping, 1065,27
thrix, hair, 1058,25
thumos, emotion, 1070,1; 1071,20
tithenai, posit, 1044,21; 1051,18*; 1084,12; 1097,11; 1104,3; 1112,14
put, 1041,18; 1104,26.27
assign, 1101,24; 1107,22*
toigaroun, hence, 1064,6
topos, place, 1042,15 et passim
trigônon, triangle, 1063,8.10

trophê, food, 1050,16
nutrition, 1076,25; 1077,13*; 1078,29
nurture, 1077,21*
tropos, way, 1049,9; 1053,13.19.22; 1060,17
type, 1049,15; 1054,5; 1081,16
manner, 1046,15
tukhê, chance, 1038,28; 1096,29; 1114,25
tunkhanein, receive, 1037,8; 1045,13
happen, 1044,12; 1063,4; 1069,21*; 1072,4*; 1098,4; 1109,32
hit upon, 1041,22
ei etukhe, if it so happens, 1111,23*
ei tukhoi, perchance, 1049,27; 1050,6; 1094,11; 1100,14.18; 1109,24
hôs etukhe, at random, 1073,2
to tukhon, a (any) chance thing, 1090,16 bis; 1096,2* bis; any chance, 1107,14
xêrainesthai, be dried, 1059,16; 1060,27
xêros, dry, 1069,21*; 1080,15
xêrotês, dryness, 1058,2; 1067,13.24; 1081,22.24
xulinos, wooden, 1062,28
xulon, log, 1055,8; 1056,5*
wood, 1062,27-1063,12 passim; 1081,21

zêtein, inquire, inquire about, 1053,13 et passim
seek, 1091,27; 1099,3; 1109,24.27
investigate, 1082,7.26; 1083,1.3; 1095,12-1096,9 passim; 1098,20
zêtêsis, inquiry, 1091,23; 1098,19; 1110,29
zôê, life, 1066,27; 1077,22*
zôion, animal, 1038,7 et passim

Subject Index